Fieldwork Handbook

FIELDWORK HANDBOOK

A Practical Guide on the Go

Marika Vertzonis

Esri Press
REDLANDS | CALIFORNIA

Esri Press, 380 New York Street, Redlands, California 92373-8100

Printed in the United States of America.

ISBN: 9781589487178
Library of Congress Control Number: 2023945926

For purchasing and distribution options (both domestic and international), please visit esripress.esri.com.

Contents

1. Introduction

2. Preparation

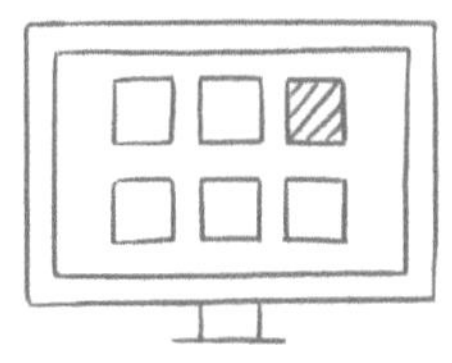

3. Equipment

4. Data

Acknowledgments

Thank you, Melanie Nash, Jim Moore, and Jeff Shaner for helping me deconstruct and reassemble countless concepts and stories and steering the eventual format of this book. Those first drafts that you reviewed look a little different now, but they are all in here, and your support and edits are greatly appreciated.

Thank you, Claudia Naber and Jenefer Shute of Esri Press, for running with this out-of-the-ordinary idea, keeping me on track to completion, and providing continual encouragement that there was an audience for this work.

Thank you, Jim Moore, Brett Stokes, Philip Wilson, Barbara Webster, Jane Darbyshire, Alix Power, Gordon Delap, and Justin Colville for sharing your personal stories from the field. While we continue to make the best possible field GIS software together, I do believe it is these firsthand experiences that best help us relate to our customers and their fieldwork needs.

Thank you, Josh Clifford, for your creative storytelling, informed content research and review, and moral support, which was invaluable to get this manuscript over the finish line.

Thank you, Carolyn Schatz and Victoria Roberts of Esri Press, for taking on the editing and design of this unique book and bringing this vision to life—that is, creating a bridge between the modern mobile computing device and a traditional paper field notebook.

—Marika Vertzonis
September 2023

How to use this book

This book is intended to be a companion for current fieldworkers or a guide for new ones. Whether you are an engineer managing a power transmission network, a scientist monitoring the repopulation of a native species, or an anthropologist studying the social impact of coal mine closures on small rural towns around the world, fieldwork is a critical element of your work. A geographic information system (GIS) is a fundamental tool in your fieldwork toolkit. Documenting the relationship between place, information, and time will help you manage your workload, allow you to make decisions, and preserve knowledge gained in the field for further research. At first glance, you may think a GIS app on your smartphone is all you need to do your job. But to ensure success in the field, it is important to take stock of the physical and digital environment you will be working in, the equipment you will need, and the data preparation you should do before stepping out the door.

The information and stories in this book come from people who build ArcGIS® apps for fieldwork. Before their move into app development, they spent time in the field in a wide range of industries. They carried computing equipment to remote locations; they dropped, lost, and damaged equipment; and they mashed keyboard keys and madly tapped on screens to get things to work, just like you. Along with the stories and feedback that customers share with Esri® every day, the developers take these personal experiences into account with every new app feature they create.

Field GIS activities are not unique to a specific industry: data capture, data validation, inspection, and field awareness are styles of fieldwork conducted in many industries. The stories in this book span many industries, and our hope is that you can draw inspiration for your own work from this broad range of experiences.

This handbook is part instruction, part story, and part activity. The information sections provide background and tips for elements of fieldwork that might be new to readers, drawing on examples from various industries. The stories offer entertainment and moral support, showing readers and fieldworkers that they are not alone, that others have been there before them. And the activities are one step removed from a blank notebook—they are guided note-taking, for a hands-on approach to planning and organizing fieldwork.

Don't go out in the field without your *Fieldwork Handbook*!

1
Introduction

What is fieldwork?

The most popular images of fieldwork in mainstream media are those of animal study. Many viewers have seen a David Attenborough TV special or video footage of Jane Goodall observing chimpanzees. Even now, whenever Jane Goodall is interviewed or giving a talk, among the images shown to introduce her are those of her in the wild, holding a camera or notebook, quietly and calmly watching chimps. These are classic images of a fieldworker doing fieldwork and are as relevant today as they were more than 60 years ago when a young Jane Goodall first stepped into the forests of Tanzania.

Consider another scenario: a person walking alongside a stream holding a clipboard, writing descriptions and the measured location and size of each structure along the bank of the waterway. With photos taken of each structure, the pages of information collected in the field are used to build a digital model of the stream that can then be used to predict behavior at times of heavy rainfall and possible flooding.

What about a team of people standing at the multiple entrances of a single train station, counting the number of people entering and exiting at 10-minute intervals? When the information from all team members is brought together, it paints a clear picture of the usage pattern of the station. It shows which entrances and exits are most heavily used, and thus where improvement efforts can be focused. In this scenario, our fieldworkers aren't moving around and carrying heavy equipment, but they are collecting data, and keeping a record of time intervals is a critical part of this fieldwork.

Now consider someone spending a month visiting one small rural village, watching the interactions of a cross section of the community and recording interviews with selected residents. In this case, interview transcripts form the large body of field data capture. Photographs and tables of data are not as important in this situation as the audio recordings and conversation transcripts.

In recent decades, the rolling news cycle has brought stark images of the range of fieldwork that forms the response to natural disasters—fires, hurricanes, earthquakes—into our homes and onto our screens. We are shown weather reporters being blown about by high winds, emergency service personnel shifting debris to locate survivors, and cleanup teams sweeping away the remains of destroyed structures and trees. As the dramatic event passes and emergency activities subside, media images may dwindle, but fieldwork activities expand. Residents, power and telecommunication utilities, governments, insurance companies, builders, and many others require accurate and up-to-date information about what was in place before the disaster and after. Existing records must be validated and new information recorded to proceed with the tasks of rebuilding and renewal.

In each of these examples, the defined geographic extent of fieldwork ranges from the wide expanses of a forest in Tanzania, to a long narrow strip of environment along the length of a stream, to a small rural village, a city, and just a single train station. They are all fieldwork scenarios, and a GIS is a critical component of the project information management for each scenario.

Drilling, fried chicken, and fishing

NAME Brett — ROLE Geologist
INDUSTRY Mineral exploration — CIRCA 2010

Before focusing on GIS as a professional specialty, I worked as a geologist in mining exploration in the desert, far from the nearest town. In one job, the task at hand was the drilling of water bores to a depth of 400 m. Working alongside the drillers, I studied and documented the contents of the spoil piles that were raised from each borehole at 5 m intervals, ready to design the construction of the water bores themselves.

I had been working in this challenging environment for months. It was hot, dirty work with plenty of waiting around while the very large and very slow drill did its work. One day I arrived at the drill rig for an early morning start and was told the drill was bogged in the mud, drill bits lost, and holes ruined. There was going to be no drilling today while the damage was repaired. Sigh. What should I do? What would you do? At this point, a nine-hour round-trip drive to the closest fast-food restaurant sounds totally reasonable, right? Whether it is or not is not to be judged here, but that was exactly what I did! The repair work took three days to complete, so the fast-food trip ended up being a fraction of the downtime but did become a core memory of that time in the field.

Another strong fieldwork memory, with perhaps a little broader cultural gravitas, is getting to know the locals. Being friendly with the nearby café owner, talking with community leaders, or simply introducing yourself to the neighbors of your project site can provide rich insight that can't be gleaned from an internet search before you arrive. It also can make the hard work you are performing in the field enjoyable.

You don't need to be doing an ethnographic study to learn from, and enjoy, time with the locals. On a different project, this time near a small village on the coast with routine downtime each day, I regularly crossed paths with a group of local fishermen who worked in the shallow ocean waters with their hand nets, catching lunch. Starting with a hello and over time working up to broader conversation, I soon found myself in the water fishing with the locals, cooking on the beach, and sharing a meal with my new friends during the project's downtime.

So, fieldwork is a whole lot more than the task you are sent out to complete.

What is GIS?

A GIS is designed to capture, store, manipulate, analyze, manage, and present all types of spatial data.

Individuals and organizations all over the world create and use GIS to model the world we live in. This geographic approach is a way of thinking and problem-solving that integrates geographic science and information so that we can make decisions, predict outcomes, and design new and improved ways to interact with our world.

By pairing location information (for example, a geographic coordinate) with tabular—or attribute—data (for example, an observation made at a nominated point in time), we can use these GIS records to visualize complex relationships on a map.

Maps can be topographic—a literal translation of the physical features on the ground—or thematic—colored or stylized representations of physical features that express some characteristic of that feature.

On a topographic map, roads might be symbolized to represent the type of road: gray lines for highways, white lines for main roads, and dashed lines for unpaved roads. The roads really aren't physically gray, white, or dashed, of course, but this symbology attempts to differentiate the physical attributes of the road in context with the greater landscape.

On a thematically styled map, roads could be symbolized according to the number of cars that use the road daily: green for low usage, yellow for medium, and red for high. A map using this kind of color styling may be useful when attempting to convey where high traffic flow occurs. Add a time dimension to this traffic symbolization, and voilà, you have the traffic layers of a vehicle navigation app.

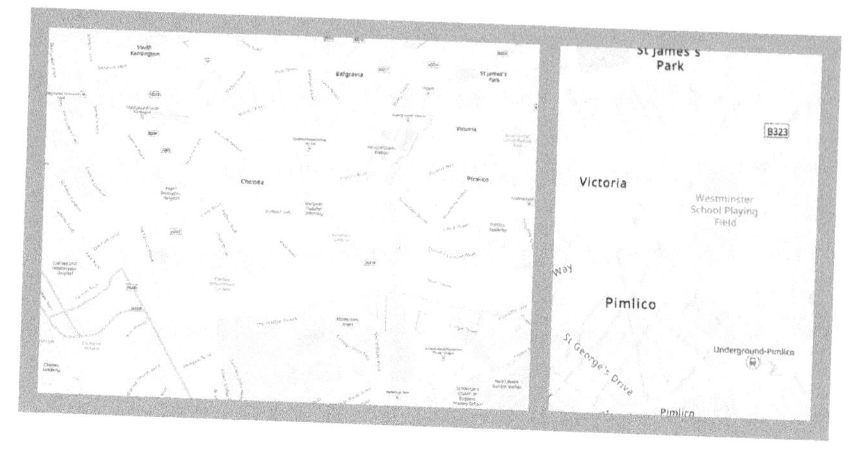

Topographic map of central London.

Most of us use GIS every day without knowing it. Popular map apps might be the easiest to recognize, but anyone referring to a timetable for catching public transportation, looking online for the closest hardware store, having garbage bins collected, or having a package delivered has had a close encounter with a GIS.

There is an application of GIS for every industry. Large-scale examples such as environmental resource management or infrastructure maintenance—pipeline, road, and rail—may be the easiest to visualize. Add to those examples retail businesses that use GIS to position new stores or choose which products to sell in which demographic areas and when to move staff from store to store. Other examples may include search and rescue teams combing an area for a lost child or firefighters determining the best way to approach and bring a wildfire under control. Consider a health department that monitors instances of disease to determine trends, causes, and required precautionary measures for a city, state, or country.

All these examples describe activities occurring in the physical world, and with a carefully constructed digital model of that world, many tasks can be visualized and planned from an office computer, a command center, or even someone's smartphone. A GIS is that digital model.

Traffic map of central London.

The physical world constantly changes, so the models must also continually evolve. Measurement tools such as high-precision GPS, laser rangefinders, and cameras are essential to update existing records and capture new ones for a GIS. The pervasive availability of personal computers, smartphones, and good internet connections means that citizen science projects can facilitate large-scale GIS data improvement exercises—from national bird count events to the gamification of satellite imagery analysis to identifying building types across large rural expanses to taking the invitation to capture (and submit) photos of native plant species thriving in local communities.

A GIS is intrinsically connected to the world that it models. Fieldwork is required to ensure that the GIS accurately models the real world while at the same time the information in the GIS guides the fieldwork to be conducted. For example, when you create a GIS to model the native habitat of orangutans in Borneo, fieldwork is essential to determine the extent of each type of land use in the study area and where injured or orphaned orangutans have been found. The GIS in turn reveals where further fieldwork should be conducted to determine the best location for releasing rehabilitated orangutans back into the wild.

What is a field GIS?

In today's mostly connected world, the expectation of having all the information about everything at your fingertips feels like a given. In practice, local data storage is finite, and internet connectivity is not always available.

The GIS used by your organization may be administered by one team, but they may pull data together seamlessly that comes from different teams and different systems in your organization for you to use. A corporate database of another design (perhaps without maps) may be critical for some information you are after, and data from completely different organizations may be needed for your field project.

Sitting at your computer in an office, these data delineations may not be visible to you—your network connectivity and security privileges may allow you to see all the data you need from within one system, and where any of that data is hosted or managed may be invisible to you.

However, when you leave the comfort and connectivity of your office, these boundaries may suddenly become apparent. Not all mobile computing devices can run the same apps, security systems can differ on different operating systems, and you simply might be forced to work offline. Along with the fact that sometimes you need only specific information in the field, the GIS you use in the field will probably be a subset of the GIS you use in the office.

When you are the custodian of all the data you need or want for your fieldwork project, there can be great benefits in taking it all into the field. Making informed decisions based on historical information can minimize the time and costs associated with fieldwork. However, having multiple years' worth of information display in your field GIS could make for a cluttered and overwhelming map. The layers, symbology, and labeling that you use in the office on a desktop app or a web map are not always suitable for use in the field.

Think of it this way: When you're packing a backpack, you want to make sure you have everything you need without creating too heavy a load. It's important to focus on taking the essentials so you can easily access the necessary tools and not have them lost in the clutter. In the same way, you want to prepare your field GIS with only the essentials that you need to be successful with your specific project.

Other sections in this book detail hardware and software considerations for hosting your field GIS, but here let's focus on the most visual element—the map. A map that shows points collected each year for 20 years can make for a busy map. Graduating the size of those points (largest for most recent and smallest for the oldest records) would help your fieldworker quickly see which ones may be most relevant today. Viewing a list of layers in the GIS and being able to turn visibility off and on should be a standard feature of any mapping app you consider for fieldwork but having one (big) button to tap that will toggle the most common layers on and off can be a game changer.

The GIS you access in the field will most likely be a subset or an amalgamation of numerous office-based systems. Your field GIS should allow you to reference the information you need to do the job in the field, but it doesn't need to have everything. Ensuring that your field GIS synchronizes with the right systems in your organization (and other organizations) is critical and will be a significant factor in choosing the right field device and app for the job.

Just like Google Maps but different

NAME Anyone | ROLE Field GIS specialist
INDUSTRY Any industry | CIRCA Anytime

When asked by family members or friends, "What do you do for work?" I tend to take a deep breath before I answer. Over the years, I've tried different ways to respond to this question, with varying success. The best answers and discussions have followed when I know what that person does for a living. In such cases, I will lead the conversation with the vague "I build mapping apps," but before their eyes glaze over, I quickly follow up with an example that is relevant to their work.

For those who were asking out of politeness, they hear enough information to remember that I do something related to their work but in IT. Those who know a little about digital maps will often respond with, "Is it like Google Maps?" Sometimes it's easiest to just say yes and move on to something else. But if I'm settling in for a long chat, a reply that's relevant to that person's own job can be a great starting point for an explanation and an interesting discussion.

I know this conversational quandary is not unique to me. Every time I go to a GIS conference—a utopian place where everybody is as excited as I am about not needing to explain themselves—I compare notes about these kinds of conversations with other attendees. We know that we all do cool things with maps—it's just that it can be hard to paraphrase for a social conversation.

We've all come to similar conclusions—that the best approach is to be armed with some relatable or recognizable examples, such as the following:

- Farmers can use a GPS and an automated chemical delivery system driven by GIS and mounted to their tractor to apply the ideal amount of fertilizer suited to each individual plant as they zigzag their way across vast crop sites.

- Plumbers repairing water pipes can see on a map what is expected to be in the ground where they are about to dig.
- Ecologists can record the location and information about species they are studying.
- Transportation engineers can understand how many people are using different forms of transport at what times and adjust timetables and resources to suit the needs of the population.

Once you start talking about a topic that is familiar to your audience (farming, plumbing, science, transportation), it's easier to describe technology in terms that are used in that field. In this way, folks begin to realize how mapping can be applied to a wide variety of everyday field tasks.

Types of fieldwork

For many people, the term *fieldwork* is used to describe record taking that is done outside an office. The information needed to perform their regular work is collected from fieldwork and used to design, improve, maintain, and inform future actions about a place or system. However, the field may be someone's regular office and fieldwork the primary form of work they undertake. When your daily routine depends on a campfire for cooking, solar panels for battery charging, and a rowboat for getting to the next worksite, your perspective on fieldwork may be different from that of the office-based worker who occasionally goes into the field. Yet your field GIS needs and goals are much the same.

The most common task usually pictured as fieldwork is new data collection or capture—walk along a footpath and capture a point that represents each tree along the side of the footpath—or inspection—walk along that same footpath, and for every point on the map representing a tree, add a note on whether that tree is healthy or needs watering.

Equally important are the scenarios of data validation and awareness. In the case of validation, imagine that someone has already digitized the trees along the footpath from an aerial image and added the species of each tree from the original planting records. You are tasked to walk along the footpath and validate that information, moving the points to a more accurate place for each tree and editing the species information if required.

In the case of awareness, you are walking along the path and just want to be able to reference the species name of each tree along the way. In this case, your own fieldwork project may not be about the tree at all. You may be an engineer about to dig up the adjacent footpath, but knowing the species of tree right next to your job may give you clues about the hole you're about to dig yourself into!

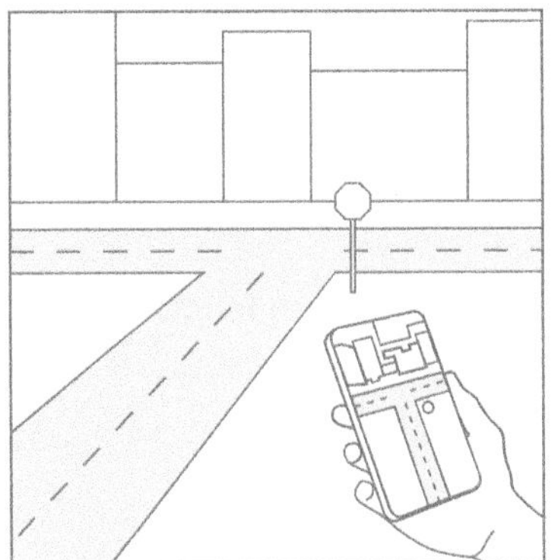

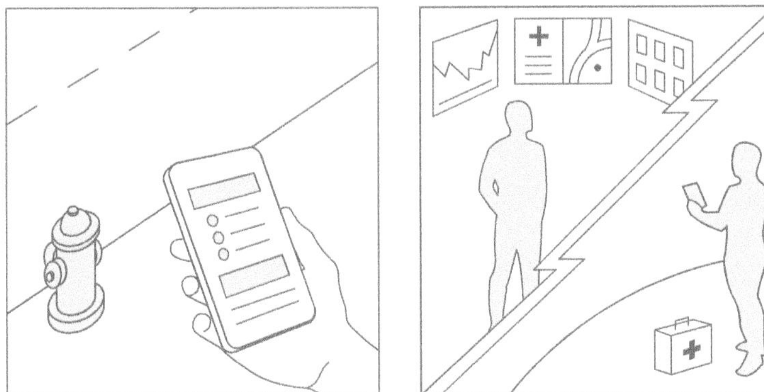

The four types of fieldwork involve different tasks.

Your fieldwork project may include more than one type of work, and sometimes you may not know it until after it's happened.

Even though we are in the field to do work, much of what we remember will be the activities and adventures around the specific work tasks. Downtime is inevitable, whether it's waiting hours, if not days, for a drilling rig to get to the next sampling depth, taking water quality samples at hourly increments, or waiting 10 minutes for a global navigation satellite system (GNSS) receiver to "settle" to capture a precise location. How you use your downtime can not only amplify your fieldwork, unexpectedly adding a different type, but also create lifelong memories. Sketching or photographing the wildlife and environment around your study site may contribute to the field data collected for what was mainly a validation project. You may not even know that the images you capture are going to be useful until months or years later. In any case, the simple fact that it filled your time and engaged your interest between measurements may be reward enough.

What's your style?

What type of fieldwork can your projects be described as? Not all projects are the same. Use these boxes to categorize your projects. Knowing the type of fieldwork you will be doing can help you plan what hardware, software, data, and other equipment you will need.

Capture

Validation

Inspection

Awareness

Capture

With minimal background information—perhaps a generic topographic map or an aerial image—fieldworkers capturing new spatial data have a clear task ahead of them: walk (ride, drive, or paddle) to the location of the feature they need to document, tap on the map, fill in some attribute information, and move to the next one. Simple, right? Every best-laid data capture plan starts like this, and then often nature (or the built environment) gets in the way.

Imagine that the row of trees forming a windbreak that you are documenting continues into an area that is fenced off, and you can't move beyond the fence line to stand next to each tree. How do you capture the trees beyond the fence?

Perhaps you are tasked to walk a line (a transect) across a paddock and capture observations of a particular plant species as individual points and, at the same time, capture the line that you walked to show which areas were traversed.

Or picture a situation in which, as you capture lines that represent sections of rail tracks that need repair, they are automatically being attributed with data from the polygon feature (the property boundary) within which they are contained.

For any of these data capture tasks, you will need an app that can go beyond putting dots on a map. You will need to capture multiple features at once or integrate the use of an external sensor or tool. You may need a laser rangefinder to capture those trees through the fence line, an app that can capture a tracklog while you capture points, or an app that can perform spatial queries on the fly while you are in the field.

The types of projects that involve completely new data capture are usually scientific research projects or projects that stretch beyond a person's or a team's normal area of responsibility when collaboration with other responsible parties (to view and use their data) is not possible.

Studying an environment for the presence of a species will result in data capture detail that may go far beyond what can be gleaned from just an aerial photo—even if the aerial photo was a significant contributor to the decision to study this portion of the field in the first place. Once in the field, new data capture of possibly significant habitat, life-form observations, or potential threats would be recorded from scratch.

Another common new data capture scenario is for impact assessments for new infrastructure. When a telecommunications or transportation corridor or pipeline is to be created for new infrastructure, an environmental assessment (sometimes several, by different stakeholders) is required. New observations are made at planned intervals and locations to quantify and qualify the impact that the new infrastructure will have on the area.

Archaeological monitoring and surveying in the construction industry can be similar. Before construction begins, a methodical survey of the full area is completed, with an archaeologist walking along transects to ensure that every square foot of the project area is sighted. During construction, observations are made as part of a watching brief that checks for anything that may be an archaeological find. As digging or ground clearing proceeds, anything of note that is surfaced must be systematically cataloged. Construction may be stopped temporarily as further investigation is done.

New data capture may be the first type of fieldwork that comes to mind, but it can be the most complex to manage or define. With a blank sheet of paper, you can draw or write anything in any direction, shape, or size. Using a printed table to capture new information can greatly improve the reliability and consistency of the records collected, but it's still useful to leave space for unplanned information. Similarly, when you are capturing new data in a field GIS, it's useful to set up data fields, default values, and automated calculations to improve reliability and consistency—but be sure to include somewhere to capture information that you didn't anticipate.

A hammer is not a stylus

NAME Barbara ROLE Archaeologist
INDUSTRY Engineering and construction CIRCA 2015

Working alongside a rattlesnake may be a nightmare for some, but for my team of archaeology and paleontology surveyors, it's just another weekday. Like our reptilian friend, my team and I could spend days in the hot desert sun, where the air is dry, and the temperature can be well over 100 degrees.

Our task was to survey planned construction sites before the ground was broken. We ensured that every square meter of land was accounted for and that any archaeological (or paleontological) resources within the project area were identified and inventoried. For example, we once found prehistoric pictographs in the Mojave Desert in California. Historic wonders, such as this rock art, may otherwise be damaged or destroyed by a new building or pipeline if not for our research.

We used mobile apps and GIS to survey these large swaths of desert land. Inch by inch, we archaeologists walked along our assigned transects, visually combing through the sands and cataloging anything we discovered. When we found something of significance, we used our app to add a point to the map that's shared across all our devices. Once the data was submitted, we could all view the new find and its location and build the picture of the project site.

Although working with a phone or tablet alleviates the headache and error caused by clipboards and paper forms, there are other factors to consider, especially when working in extreme climates. Sometimes our devices would overheat, or the glare of the sun would interfere with our ability to see the map we were working on. Not to mention, the devices themselves are a lot more fragile than your average clipboard.

Working within the same project area, paleontologists would focus on fossilized finds and, adding to their standard fieldwork uniform, would hang a rock hammer from their belt. Convenient tools are all good, until you also have a computer tablet "carefully" housed in a rugged-edged casing with a strap attached so that it can be slung across your shoulder for easy carriage. Always take a moment to be sure it's not slung on the same side of your body as the rock hammer! Digital screens, hammers, and movement (as you scramble over hot, rocky terrain) don't play nicely together.

Human error aside, with mobile devices in hand, we're able to locate, inventory, and preserve some true marvels.

Planning data capture

Even when you plan to capture completely new information, you can, and should, prepare material to help you. Whether it is a basemap or complementary data points, whatever you can take with you can help speed and improve the quality of data capture. However well you plan, you may still discover new information you want to capture while in the field. Use the following prompts to identify what materials you need to prepare.

Basemap

Aerial imagery | Street map | Topographic map

Structured data capture

ID | Name | Address | Observation date | Size | Condition

Complementary data

Roads | Property boundaries | Building footprints | Trees | Fences

Plan B

Generic data layers | Paper notebook | Printed maps

Notes

Validation

Over the last 20 years, our ability to store, convert, and design fieldwork GIS projects has radically evolved. Gone are the days of going into the field with a blank screen and capturing everything from scratch. This is a time-consuming, error-prone, and expensive task. By starting with a digital basemap and features that have been digitized from aerial imagery or derived from as-built drawings, a fieldwork project can be streamlined and become a validation exercise rather than a raw data capture exercise. This situation typically means that it can be done more quickly and more efficiently, which ultimately means more cheaply, than when starting with nothing.

The irony of a GIS as an information model is that you need to know what you're going to put in it before you put it in so that you can design the places for it to go in the database. This is not unusual; any type of database is the same. And frankly, any storage project is the same. How big a box do I need for all my winter clothes to make space in the closet? Do I use two small boxes and split them up into tops and bottoms? Do I need another box for shoes? If I split them up, will it be easier to retrieve just the shoes if I need them?

A primary goal of asset management GIS is to replicate the real world as closely as possible, so that when you use it to model actions or create predictions, the GIS can demonstrate what would happen in the real world.

Typically, a GIS of this nature would be constructed from the initial design data—digital drawings and designs from architects and engineers showing how the real-world system was built. Pages and pages of drawings with detailed minutiae are created for construction, but whether every one of these is updated after construction to form a set of as-built drawings is another story. In practice, as-built drawings can take the form of a design drawing with a bit of red pen scribble indicating differences. Too often, a GIS represents what the system should look like, not what it really does.

In the field, validation is critical. Should you give every field technician a phone or handheld computer to record all the information? Yes, no, maybe. If all they are recording is the progress of a current job, then perhaps yes. Tick a box and fill in a short form to indicate progress. But if you are looking to fill in pieces of a greater asset management puzzle, consider having a dedicated fieldworker on-site to capture all the information.

Even if you cannot have a dedicated person join the field crew to capture data as the team works in the field, it's worth taking time to choose the most appropriate data-recording tool for the job. If your work relies on recording a measurement and then performing a calculation before your next field task, set your device and app of choice to display the results of the calculations to you in real time. If your team needs to see all the information about what has been recorded for this site, organizing the information on a single screen or well-ordered set of pages will ensure that your fieldworkers complete their work most efficiently.

Carefully organizing the screens of information in an order that matches the fieldwork workflow or using a device that has a screen big enough to show all the information at once can mean the difference between data capture success or failure in the field.

Wastewater treatment plants don't smell bad

NAME Marika ROLE Graduate engineer CIRCA 1998
INDUSTRY Waste and wastewater management

My first job out of university was in a small-country local government area (LGA). The LGA managed its own sewer and water systems for two towns, one servicing 10,000 people, the other servicing about 3,000 people. There was a small engineering team that maintained the network, replaced and repaired pipes when damaged, installed new portions of the network for new developments, and kept the network operating smoothly. They also maintained two small sewage treatment plants. The plants were textbook-style installations, with input filtration, trickling beds, and settling ponds.

I distinctly remember going to visit the smaller of the two treatment sites. The first thing that struck me was that it didn't smell as bad as I thought it would, and second, it was all much cleaner than I expected. The entry point was the most striking. At that first stage of processing, you see things that shouldn't have gone in the sewer in the first place—typically plastics and rubbish. Once these are screened away, the things that can be treated are churned up, biologically chomped, and returned to nature in various forms. Treated water was keeping the local golf course extra green, and the relatively small amount of solid waste was applied to suitable land locations as fertilizer.

The head plumber and his team of three knew every part of the system from memory. There was even a book of drawings of the network. Fifty or so printed A3 pages were bound together to represent all water and sewer pipes in the two towns, the pumps that extracted water from the river, and the two treatment plants. There were scribbles that indicated new pipes or replacements and notes that described what had been done. Some were dated. Some were not. Sometimes materials were listed. But mostly, all that detail was stored in the mind of the head plumber.

The mind is a superb relational database, one that is difficult to replicate digitally, but it doesn't serve well when its host would like a holiday or new team members are brought on board. My job on this team was to extract as much information from the book and the head plumber as possible and store it in a way that was readily accessible and updatable. There were some digital files for those drawings—it shouldn't take too long, right?

The first task in transforming the paper plans for the sewer and water systems was to list all the things that were represented on them: trunk mains, distribution pipes, risers, valves, connectors, pits, and pump stations. Some locations had a lot of things represented in a small space. And the same type of structure came in different sizes, material types, and technical variations. All combinations needed to be represented by their distinctive features. That way you could search for all clay pipes, all three-inch pipes, all brass valves, or all brass valves installed before 1995 in the area north of the showgrounds.

With the data model in place, the next task was to get all the information into the new database. With varying data formats and different levels of confidence in the available information, this was a divide-and-conquer effort. Starting with the most basic but most extensive information, I reformatted, massaged, and imported whatever I could. Most of the effort was spent refining and adding detail to newly minted records and creating new records that better represented the real-world system.

Because this was a small town, the plumbing team was never too far away, so when a trench was dug for repairs or replacement, the opportunity for the most important task—field validation—had to be seized. No amount of sifting through paper service reports to determine whether a pipe is cast iron or clay can replace seeing the pipe itself.

Even though I have gone on to see and use GIS in a broad range of industries and applications, this first data conversion project that I undertook forms a clear picture in my mind of what a GIS is—a digital twin of a physical system and the reason why validation with fieldwork is critical.

Park bench condition assessment

Asking fieldworkers to validate or refine existing data is an important way to set guidelines and boundaries on what information is recorded during fieldwork and leads to the creation of consistent and methodical information models.

Example assessments

If you asked four people to go out and record the location and describe the condition of the same three benches in the nearby park, the information they each record may differ in geographic coordinate system, naming conventions, and level of detail. Consider the following four example fieldworker reports for a park bench assessment.

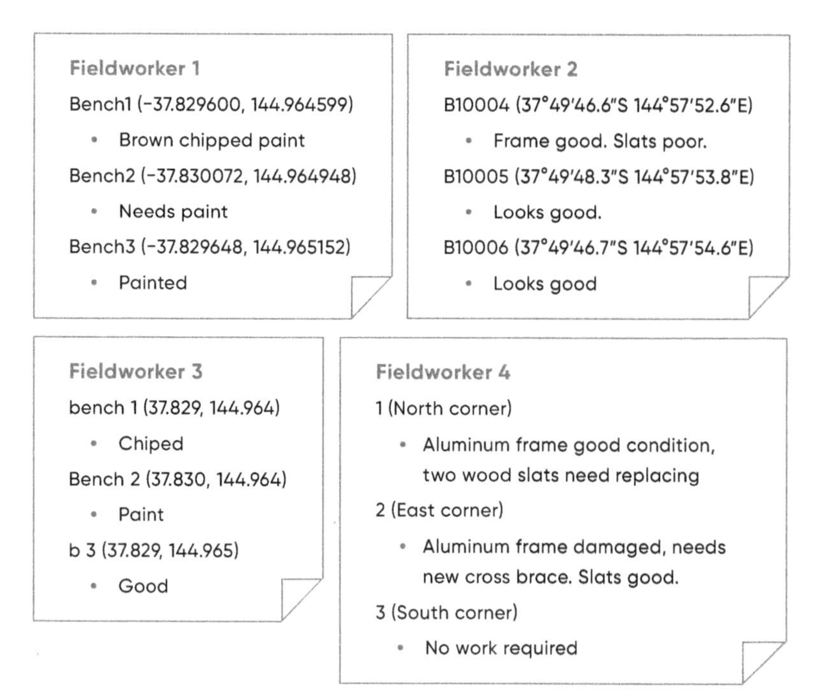

First, each record in the example assessments is identified by a different set of names. Fieldworker 1 has used a unique name for each bench with consistent capitalization and spacing in the name. Fieldworker 2 is either quite creative or knows something the others don't; they have used a naming convention that looks like an asset numbering system. Fieldworker 3 doesn't have the patience or interest in a naming convention—they are just typing stuff. The universe might just be revolving around fieldworker 4: the numbers 1, 2, 3 are all they use.

Next, the location information for each bench differs significantly. Fieldworker 1 has used geographic coordinates, listing the latitude and longitude of each bench. Fieldworker 2 has listed easting and northing coordinates. Fieldworker 3 appears to have been rushed: not only have they dropped some important decimal places, losing at least 100 m in accuracy, but they have also lost the minus sign, moving our park bench into the ocean in the wrong hemisphere! Fieldworker 4 used a descriptive location, which, without a clearer requirements statement, you really can't fault. They have described the location of the bench relative to the park itself.

Last, we see the condition descriptions. Each one tells a different story. Fieldworker 1 was most interested in the paintwork, fieldworker 2 described the benches, fieldworker 3 wrote in shorthand (is the second bench painted, or does it need painting?), with a typo to boot, and fieldworker 4 is already budgeting the repair work.

None of these field reports is wrong, but if you had sent these people to four different parks, it would be hard to combine their records to create a single list of repairs that needed to be completed, let alone create a park asset inventory for ongoing asset maintenance.

Alternatively, if each of these fieldworkers were given a digital map with points already identifying the benches—and when they clicked on each of the point locations they could choose a condition description from a list—the typos, shorthand, and general storytelling could be eliminated from the source.

Add an automated username and date and time stamp to the park bench inspection record, and you have the makings of an asset management database that can be used to alert maintenance staff to benches that need repair. It also has all the details needed to know when a bench was last inspected and by whom, and to plan when it should be inspected again.

Design a park bench database

Use the following chart to design a database suitable for capturing condition assessment information for park benches. Some field names have been added to get you started.

Field type (string, integer, list)	**Field name**	**Choice list values (if applicable)**
Easting, northing	Location	n/a
	AssetID	
	PaintCondition	
	StructuralCondition	
	InspectionDate	
	FieldworkerName	

Notes

Inspection and monitoring

Inspection and *monitoring* are terms that describe similar actions, each preferred by different industries. In the built environment, we tend to inspect facilities and features that we maintain. In the natural sciences and agriculture, we monitor systems and features. In both cases, we are reviewing information that already exists, updating it, and appending it.

Many countries have a "toward zero" philosophy regarding hazardous incidents on a worksite, and to ensure this, regimens of inspection, checking, and testing are a staple in operating any infrastructure or equipment that can cause harm. Maintenance and inspection records are a critical part of keeping complex systems working. Even without the human safety consideration, infrastructure costs money, and it costs even more when it's not operating.

Consider how this kind of safety inspection would take place when the critical piece of infrastructure is a gas pipeline that extends for thousands of kilometers and crosses state or national borders. Water, sewer, stormwater, or oil pipelines are no different. Roads, rail, and electrical power pole easements all occupy lengthy corridors of land that stretch across our environment, creating vast networks of infrastructure that need to be maintained.

Gas and oil pipelines traverse states, countries, and continents. When fuel stops flowing, industries, if not entire nations, can grind to a halt. The opposite of no fuel can be equally catastrophic: pipeline failures and subsequent leakage can cause unimaginable devastation to native wildlife and the surrounding environment. Accordingly, the operation of fuel pipelines comes with comprehensive compliance reporting, with daily patrols designated for hundreds, if not thousands, of kilometers of linear infrastructure.

One blessing in what could be considered an overwhelming task is that, for operational purposes, the content needed for an inspection field GIS is typically already well documented. Design and installation details used to construct the infrastructure constitute the same information needed for inspection and maintenance efforts. The infrastructure doesn't move, so there's no need for field inspectors to capture locations on every visit. Instead, a well-designed database can have workers append an inspection record quickly to the existing information about the feature they are inspecting.

When the inspection is going well, it may be no more than a date of inspection confirmation, but when a hazard is identified, more information may be collected. When covering many kilometers a day, a fieldworker inspecting infrastructure may be capturing records quickly (driving alongside the linear asset) or getting up close and personal (stopping at each significant landmark). Or they may be standing on a crane raised to the top of a power pole and studying (and servicing) the equipment attached to the top of the pole.

In contrast to vast corridors of infrastructure inspection, heritage conservation inspections are typically more localized and more detailed. Consider a heritage-listed building whose preservation management must conform to a specific and regular maintenance program. Different levels of condition may need to be reported for different aspects of the building: Are the windows in good repair? Broken glass? Rotten window frames? Do the doors close safely? Is the roof leaking? Electrical safety? Different parts of the site may be inspected by different teams, so each team will need the detail appropriate to their task.

Inspection-style fieldwork typically makes use of existing data in a field GIS as reference material only rather than requiring full editing capabilities of all records in the GIS. Providing the ability for a fieldworker to add information only (and not edit existing information) can significantly change—and typically reduce—the data access permissions and app complexity needed for the job.

Breathe in, breathe out

NAME Justin ROLE Software developer CIRCA 2005
INDUSTRY Environmental health and safety

One of the largest projects I worked on was developing an app for inventorying asbestos-containing materials in buildings across Ontario, Canada. Buildings ranged from telecommunication facilities to properties owned by the Toronto District School Board. As you may know, asbestos is a dangerous and even deadly substance—which is why it's so important to understand where it is. There are even federal regulations for asbestos as part of the Ontario Occupational Health and Safety Act. To ensure asbestos was properly inventoried across the facilities we were responsible for, our solution had to be efficient, robust, and, most importantly, accurate.

Before our team's involvement, these buildings were inspected using paper, pens, CAD drawings, and digital cameras. This presented a variety of issues, including inconsistencies in data, mislabeling of forms, illegibility, and other errors—all across inspections for hundreds of buildings.

Our solution? Creating a custom app for field data collection.

My team and I developed a mobile app that streamlined the inspection of these facilities and the inventorying of asbestos-containing materials. The app contained a form that inspectors could use to capture data for each room, including the ability to copy repeated attributes between rooms, require that certain data be entered, and conditionally show or hide parts of the form based on earlier data entry. We even incorporated bar code scanning to automatically populate the form using sample codes for each unique material being sampled for testing. Surveyors were still using digital cameras to capture photos, but that's just because the modern smartphone hadn't been invented yet.

We were also able to make use of the original CAD drawings, extracting data on general facility information, such as room names, to prepopulate the database before surveyors went into the field, further helping to streamline their data entry in the field.

In short, we completely transformed the way these building inspections were being conducted.

The result was a more efficient system for collecting data and centralizing it, all in one place. We were able to integrate new asbestos data with existing information from CAD drawings, as well as incorporate lab results from the materials themselves. Once all this came together, we were able to automate generating reports for each site—creating a streamlined way of sharing important information on asbestos-related health hazards.

The system was a success, and the app would later be expanded to collect additional hazardous material information, such as spaces that included lead-based paint. This set up a great framework for future building inspections and inventories and would allow organizations to communicate information and take action regarding the environmental health of their facilities.

Understanding inspection requirements

Inspection projects provide the most opportunity for advance planning, optimizing data capture methods for speed and accuracy, and informing next activities. Use the following prompts to identify the information you need to collect, how often, and for what kind of subsequent activity.

What information needs to be collected at each inspection?

Condition | Obstructions | Hazards | Data | Time | Inspector ID | Next actions

How often are inspections required?

Daily | Weekly | Monthly | Annually | Ad hoc

What information typically won't change at each inspection?

Location | Structure type | Make | Model | Material | Size | ID

What kind of activities can result from an inspection?

Regulatory reporting | Infrastructure maintenance | Infrastructure replacement

Awareness

Not all fieldwork is focused on collecting information about the real environment to take back to the office to analyze. Sometimes it's the other way around. A key capability of a GIS is to create a digital model of our world so we can learn, predict, improve, and maintain systems outside. Taking those digital tools with you when working in the field can help you make informed decisions while you are there.

Wildfire response teams use apps when attending wildfires to support their on-ground and in-air operations. Some people within the greater team may have data collection tools in hand, but many on the ground have other priorities.

Being able to see wind predictions and create simulations in the field can be a game changer for teams responding to wildfires and vastly improve safety and situational awareness. In search and rescue operations, the ability to see the ground (or ocean) that individual trackers, or teams of trackers, have covered is critical in planning the next search session. Overlaying the tracklogs of walkers, dogs, drones, or helicopter search teams with artifacts found can help pinpoint critical locations to perform the next more detailed—and possibly successful—search.

Situational awareness is also a primary goal in large-scale public event management. Marathons, car and motorcycle racing, and street parades have complex logistical requirements and an extremely high public profile. Both operations management and emergency response can be significantly heightened with field GIS.

In the emergency communications center (ECC) of these events, the critical objectives are all the same: Where are our people, and where do they need to be? At marathon events, emergency response team members are equipped with field devices with a cellular connection that allows the ECC to see where they are at every junction while the event is in progress. The closest available team member to an incident can be chosen with a glance at an overview map, and a radio call to the right person can have them deployed immediately.

In these types of events, the terms *command*, *control*, and *communication* are often used, but each has a distinct definition. Depending on the type and level of incident, those in control and command may be specific people and roles. The situational awareness map may be visible to all groups—operations team members, police, fire, ambulances, and first aid—but for any given incident, a different person or group may have designated command. If a barricade has been pulled down by a group of spectators, it may be the police in command with the primary objective of returning order. If a barricade has come down in a freak lightning strike and has crushed a group of spectators, it may be that the fire chief is in command. All other agencies might be involved—ambulances to tend to injured spectators, operations to clear the debris, police to keep onlookers away—but the fire chief would be ultimately in command for incident management.

No matter how advanced the available technology, someone will still ultimately be in command and need to decide. Technology, including a field GIS, is a tool that informs communication, but command and control are still clearly separate operations.

Can't take your eyes off the big (GIS) screen

NAME Marika | ROLE GIS specialist
INDUSTRY Events | CIRCA 2003

Major sporting events can command the attention of the whole world—motor racing, team sport world cups, and grand slams. But just as the spectators at the event are eagerly watching the participants strive for glory, many people are staring at a big screen (or many screens) watching the spectators. Police, ambulances, fire, doctors, nurses, and security work together to keep major events safe for everyone.

Over several years in the early 2000s, I was integrated into the emergency communications center of a Formula 1 event. I shared and maintained a GIS of the facilities used by spectators around the track; these were temporary facilities that had been built from scratch in just the previous few weeks. Seated with me around a horseshoe-shaped set of desks (purposefully designed so we could all eyeball each other at any given time) were the emergency communications manager (in this instance, the chief of police), a chief from each of the emergency services—fire, ambulance, police, first aid—and event staff.

We would all spend four days watching the big screen—the big-screen GIS—seeing minor incidents pop up: a spectator fell and scraped their knee—St John first-aid workers in attendance; a spectator suffering from heat exhaustion—an ambulance took them to the hospital; a scuffle between spectators at a food van—police broke up the incident. Each incident was assigned to an individual service or a combination of services, and resources were deployed immediately. The event venue was essentially a pop-up site—a temporary village constructed in the middle of an inner-city park. Local knowledge of the area was of little help, but with the GIS we were able to pinpoint the closest person or equipment that could be used, and then helped team members navigate to the incident.

— Who is in control, command, and coordination? —

Whether your field operation is a military exercise, a natural disaster response, or management of a major sporting event, the effective collaboration of multiple teams toward the common goal is critical. How teams work together and who the decision-makers are in different scenarios should be made clear to all from the outset. In these types of field operations, many C-words are used, often interchangeably. A clear understanding of at least three—control, command, and coordination—is critical to success.

Get to know the three critical Cs by answering the following questions. The word *team* may represent different agencies, many companies, or groups of people within a single organization.

Control

Control is a cross-team activity that guides the overall direction of the field operation. What is the objective of your field operation?

....

....

....

....

....

....

....

....

Command

Internal to each agency or team, command describes how activities are accomplished.

In the following chart, list all the agencies and teams that are involved in your field operation, what they are responsible for, and what their internal command structure is. Who is the ultimate decision-maker in each agency or team?

Team	Area of responsibility	Command structure

Coordination

Cross-agency coordination is the most critical element of a field operation. How will the teams work together toward the common goal, and who is ultimately responsible when decisions are required?

In the following cross-agency coordination diagram, enter the teams that are involved in your field operation and list the names of their key points of contact. Who will be responsible for critical communications between teams? Which team has the final say?

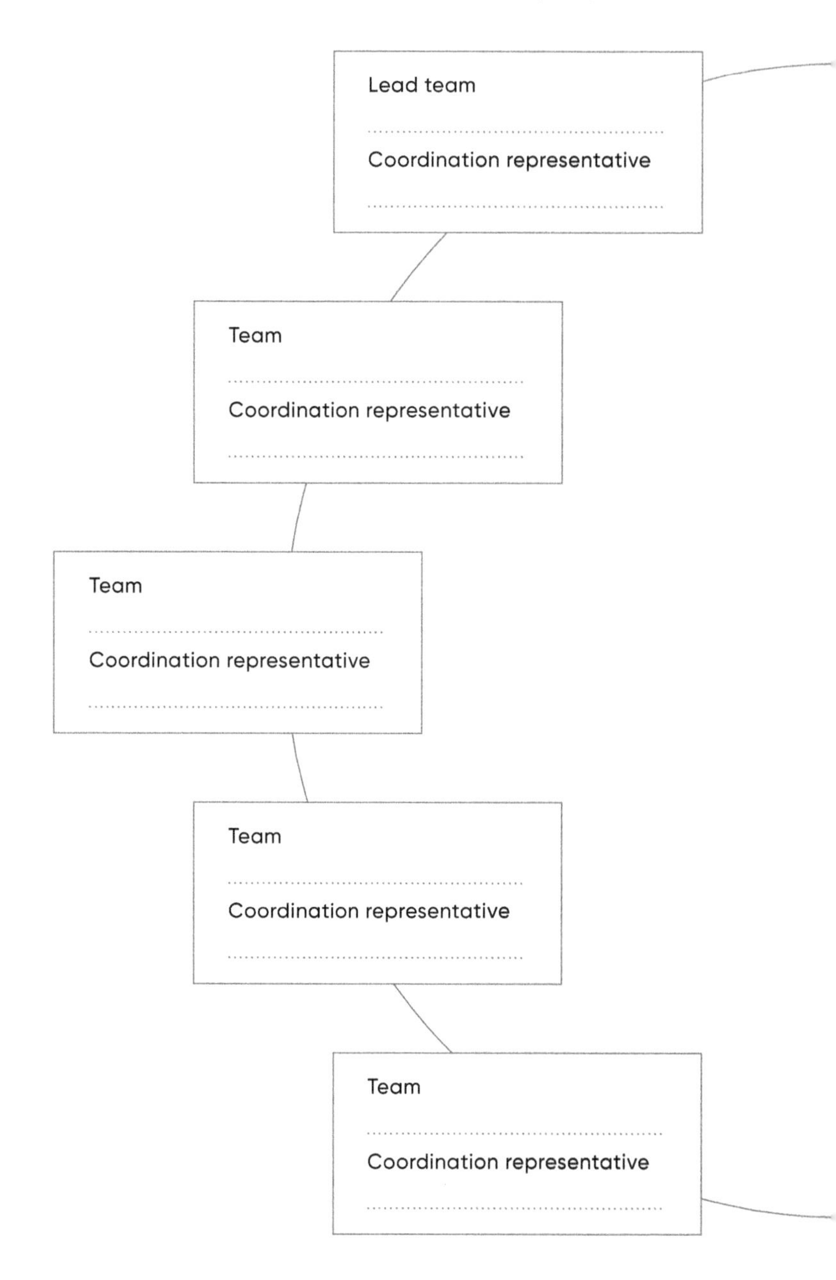
Lead team
Coordination representative
Team
Coordination representative
Team
Coordination representative
Team
Coordination representative
Team
Coordination representative

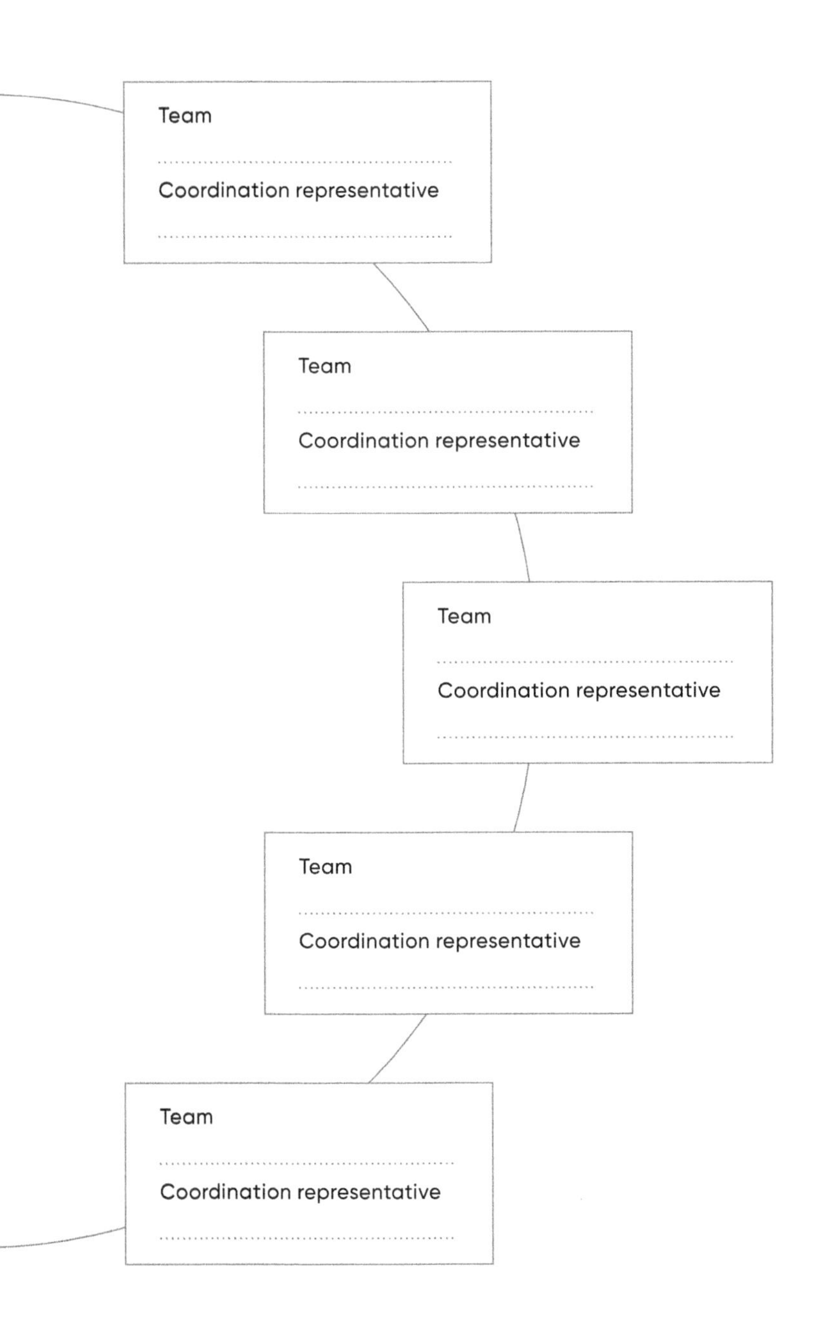
Team
Coordination representative
Team
Coordination representative
Team
Coordination representative
Team
Coordination representative
Team
Coordination representative

Notes

2 Preparation

Trial and pilot fieldwork

Designing and scheduling trial and pilot fieldwork projects is a great way to test your planned workflows and better estimate your overall project requirements. Depending on the scale of your planned fieldwork, your trial may involve just one fieldworker (you) and either geographic constraints or a subset of your time.

Consider planning a curbside heritage building assessment for a small city with 15,000 residences. Assume it takes five minutes to complete the assessment—confirm the address against existing records, fill in a form that describes the condition, take photos. A few back-of-the-envelope calculations show that the whole city could be covered by about 15 fieldworkers in two weeks. With those kinds of numbers, this project sounds easy.

Yet, to add substance to this estimate, it would be important to do a field trial. Employing 15 people for a short time, or fewer people for a longer time, represents a significant investment in people and technology. It's important to get it right.

Performing a data capture trial for, say, two hours (at different times of the day) in different localities in the city (inner urban, suburban, semirural) will go a long way toward getting a better estimate of average and peak data capture times. In inner urban areas, travel time between residences will be negligible, but on the outskirts of the city, the fieldworker may need to drive between residences. After conducting these trials, you may determine that, because of traffic, it is impossible to meet the five-minute completion time estimate; it's more like 10 minutes. Restricting the day's work hours to omit peak traffic times will also help, so more appropriate resourcing might be 10 fieldworkers for four weeks.

Don't forget to put your technology choices to the test during the pilot. Train this small group of fieldworkers with the view that when things scale up, they will form a part of your pyramid training scheme. These people can work with and train others. Encourage them to provide feedback and make suggestions for improvements to the fieldwork process. Be prepared to change or fix things on the run. While you have these early adopters in the field, test the changes and validate that they do make the fieldwork better. This will save you headaches in the long run.

Allow time for tea

NAME: Marika ROLE: GIS specialist
INDUSTRY: Population census CIRCA: 2010

Several years ago, I participated in a field trial for a population census, with a small group of test field enumerators. Our primary goal was to test the questions listed on the form with real people and in the working environment (not behind a desk in the office). But what we quickly learned was that keeping to the planned time allotment was going to be tough.

In this trial, the enumerators were members of the government department responsible for the project. They were going to experience the process that they had commissioned and get firsthand experience of what their teams would encounter in the field. A few members of the team who put together the device, app, and form of choice tagged along to observe, advise, and troubleshoot. This census was for a country in the Middle East, and the form could be viewed in English or Arabic.

We split into two teams of four and were sent to an inner urban area with mostly small apartment blocks and some single household residences. My group approached the first apartment building, and after quickly identifying the correct building in the app, morale was high.

Then the designated team leader pressed the intercom button for the first apartment. No one was home. The response to the pressing of the second button was a rushed, somewhat agitated flurry of words and a hang-up. With a little dent in our morale, we pressed the third button. After a quick explanation, we were let in and greeted kindly by a woman who was the head of the household.

We were welcomed into the living room, tea was served, and many minutes of discussion (in Arabic) ensued before an enumeration device was even touched. Discussion eventually came around to the census, and the form was completed. There were many children in the family, so the multilevel form that accommodated capturing related records for each child got a thorough workout. When testing in the office, choosing to add just one record is too easy. But if you are completing the trial honestly, you can't do that when four small children are sitting there in front of you.

Now, this was a trial, and I do suspect somebody from the team knew somebody from the family, which contributed to the chatter, but 30 minutes later when we emerged from that one household, there was silent realization in the eyes of all our team members that this fieldwork was going to take a lot longer than originally calculated.

Notes

Plan a trial

Answer the following questions to help plan your trial project. Knowing the answers to these questions can help you better group activities according to the resources—human and technological—that you have.

How many sites, features, or activities can be done in a day, week, or month?

....

....

....

How long does an activity take?

Days | Weeks | Months | Years

....

....

....

Do all activities consist of the same steps?

....

....

....

How do the activities need to be completed?

In sequence | In parallel

Who performs the activities at a specific location?

Same person | Different specialists

If fieldwork is to be broken into geographic areas, how do you define them?

Conventional street addresses | Custom-designed polygons

Risk assessment

In many industries, fieldworkers must follow risk assessment procedures in the field. On construction sites, potential risks are discussed at daily toolbox talks, and methods to avoid risk are highlighted.

Formal documentation may not be a requirement in your industry, but integrating the documentation of a fieldwork risk assessment into your field GIS is a good way to ensure your teams plan their work before they head out. Also, in case something goes wrong, you will have documentation that describes the actions taken to minimize the risk.

To assess risk, potential hazards are identified, the likelihood of occurrence and potential outcome of each hazard are rated, and depending on the outcome, mitigation measures are put in place to reduce the risk.

These ratings can be readily illustrated with a risk matrix.

To use this matrix, for each hazard identified, choose the likelihood of occurrence, and multiply the corresponding value by the value corresponding to the potential outcome. A resultant value of 15 or greater is considered a high risk, a resultant value of 5 or lower is considered a low risk, and a value in between is considered medium. Depending on the project, the organization, or requirements from external agencies, all risks may need to be mitigated to a specific level.

For example, consider the potential hazards during a fieldwork project in the Nevada desert to perform an environmental survey of the route for a proposed pipeline infrastructure project. In this project, assume that all risks must be mitigated to a level of medium. Risks in this project may include being bitten by a rattlesnake and dehydration from the hot conditions. The likelihood of being bitten by a rattlesnake may be an even chance (3), but with a potential outcome of fatality (5). The likelihood of dehydration may be

Risk matrix

Likelihood			Potential outcomes: Minor injury	Injury needing medical treatment	Injury needing 1–5 days off work	Serious injury or long-term sickness	Fatality
			1	**2**	**3**	**4**	**5**
Likelihood	Near impossible	**1**	Low	Low	Low	Low	Low
Likelihood	Unlikely	**2**	Low	Low	Medium	Medium	Medium
Likelihood	Even chance	**3**	Low	Medium	Medium	Medium	High
Likelihood	Likely	**4**	Low	Medium	Medium	High	High
Likelihood	Near certainty	**5**	Low	Medium	High	High	High

Using the risk matrix for a potential hazard, multiply the value for the likelihood of occurrence by the value of the potential outcome to determine the gravity of risk.

a near certainty (5), with a potential outcome of injury needing medical treatment (2).

With a resultant score of 15, our rattlesnake risk is too high, and mitigation measures must be put in place. Education for the fieldwork team on how to identify preferred rattlesnake habitat and how to act when a rattlesnake is spotted, along with clear directions for taking the bite victim to a hospital, may allow you to change the likelihood to unlikely (2) and the potential outcome to an injury requiring time off work (3). With these scores, the risk would be considered medium.

Our dehydration risk, with a score of 10, is already medium, but mitigation measures of hats, sunscreen, water allocations, and breaks in the shade can help your fieldworkers and readily bring the risk down to low.

It's not just people who are affected by the heat. Computers, tablets, and phones don't take well to hot environments. Working in short blocks of time and resting in the shade are good for humans, animals, and computers. Some modern smartphones, tablets, and computers will even alert you when they are overheating. Consider having a dry zone in your portable ice chest (if you have one), or, at the very least, have somewhere shady for you and your equipment to rest.

Even without making a habit of assessing risk, ensuring personal protection and safety should come naturally. If not, the environment sure has ways to remind you. Risks associated with extreme weather—heat, cold, wind, and water—can form some of the more memorable moments of fieldwork.

Working on and around water adds its own set of risks. Whether you are on a boat in a lake measuring water quality, wading through wetlands searching for a specific wildlife habitat, or just needing to continue your work through the daily afternoon rainstorms in the wet season, planning how to use technology in the wet is critical. Your average consumer smartphone is not up to the challenge of being dropped in a puddle (or lake) or having buttons pressed with wet hands. A waterproof case for an iPad may be enough, but consider mounting your device to the boat, land vehicle, or pole to minimize the chance of dropping your valuable data capture into a watery abyss.

Horseback rescue

NAME: Jane ROLE: Paleontologist intern CIRCA: 2006
INDUSTRY: Paleontology and geology

The biggest lesson I learned while doing fieldwork with Petrified Forest National Park in northeastern Arizona is how important it is to have daily safety meetings and to exercise caution every time you're going out in the field.

During my summer with the park service, my team and I were tasked with surveying fossil material throughout the Arizona desert. It was scorching hot, dry as all get-out, and there was always the occasional rattlesnake or two you had to watch out for. As you can imagine, there was a lot of risk involved, which is why our daily safety meetings were so important.

Keeping safety in mind is crucial, because it's so easy to make a small mistake that creates a big problem. Remembering to charge your radio battery, stay hydrated, and always use the buddy system are simple actions that can prevent you from landing in a less-than-ideal situation. This is coming from someone who knows firsthand how quickly a day of fieldwork can take a turn for the worse.

It was one of the hottest days on record, and we had already hiked about 14 miles. I had been drinking water like a fish (at least four liters). Unfortunately, the heat started getting to me, and I started feeling worse and worse as the day dragged on. It got bad enough that, in the end, I had to be rescued by park rangers...on horseback.

There weren't any roads out in the desert, so the only way you could get medical assistance was by radioing in and calling for a horse ranger to come pick you up. So, I got to ride a horse—but I wouldn't recommend getting heat exhaustion to do so. By the time the horse arrived, I was feeling better, but everyone assured me that I had made the right call about radioing in and requesting help. Thankfully, my radio battery was charged, and the team I was working with knew where I was located. If that hadn't been the case, I'm not sure what would have happened.

So, let people know where you're going every time you venture out in the field, always take precautions, whatever your surroundings, and work with your team to make sure you're having safety meetings at the start of each day. Those frequent reminders are necessary to ensure the safety of yourself and your entire team. And perhaps also the horse.

Perform a risk assessment

Calculate the risks associated with one of your projects, and be sure to add mitigating factors to lower the value.

Potential hazard	Mitigation measure	Outcome	Likelihood	Risk	Mitigated outcome	Mitigated likelihood	Mitigated risk
Extreme sunburn	Hat, sunscreen, long sleeves	2	5	10	1	2	2

Field office

Depending on the amount of time you will be in the field, the number of teams involved, or the proximity to your regular place of work, you may need to set up a field office.

For some people, this may be their only office; for others, it may be a hotel room intended as an overnight stopover or a campsite used for a few weeks or months. Your field office could be a customer or business partner's office, a portable building with a diesel generator for power, or a tent with a solar panel connected to a collection of car batteries and a laptop sitting on a pile of boxes.

Consider, for example, a trip to another country to do fieldwork. A visit to a local university results in collection of unexpected maps and data (and students to help with the fieldwork) that will significantly aid the pending

Any space can become a field office, even a hotel room.

fieldwork. A quick trip to the local office supply store and the purchase of a scanner and printer turns a modest hotel room into a GIS administrator's temporary workspace.

If a hotel room in town sounds a little primitive, picture the mountains, jungles, and deserts where natural habitat research, mining exploration, or greenfield infrastructure projects might take place. Your power supply may be intermittent, coming from a diesel generator or renewable sources, with equipment brought in on the back of a truck driven for many hours from civilization to a campsite. In these circumstances, planning what you will need in your field office is critical. There is no quick trip to an office supply store when your field office is in a barren desert or the jungle.

Wherever, or whatever, your field office may be, the purpose of this space to is enable you to organize and coordinate your fieldwork and fieldworkers.

Plan for everything you are going to need because sometimes your fieldwork may be done far from anywhere.

At the most basic level, this space should be dry, out of the heat, able to provide shelter for your equipment, and preferably somewhere where you can charge batteries and perform running repairs.

You must also consider security. A pop-up field camp may appear as an attractive collection of resources, ripe for poaching, to local human and animal populations alike. Securing everything from food to equipment to building materials is critical, no matter how far away you are from supposed civilization. A cabin in the jungle of Borneo used as the base for primate study will need door and window latching as sophisticated as any building in town. Orangutans are clever and curious; they could easily turn your workspace and work materials upside down if they sneak into the cabin.

Field office requirements

Transport, shelter, power, and security are the critical elements of establishing a field office.

In each box, circle the options that best describe your field office and add your own details. Keep this list in mind when identifying a location, purchasing equipment, and planning site setup and ongoing maintenance.

Transport

Truck | Car | Boat | Foot

Readily accessible | Limited accessibility

Frequent visits possible | Infrequent visits

Shelter

Roof only | Complete enclosure

Excessive heat | Fire | Cold | Rain | Wind

Power

Main power | 12-volt power

Solar | Petrol generator

Continuous | Intermittent

Security

Entire office space lockable | Equipment only

Malicious people | Curious animals

Training fieldworkers

As diligent creators of field GIS tools, you test your app, map, or form outdoors in the environment where it will be used. You carefully choose an app and device that suit the project needs. After other people use it and provide feedback, you incorporate it into the final design. But will everyone be able to use it as intended?

In the utility and construction industries, the primary role of fieldworkers is to build, maintain, or repair infrastructure. Recordkeeping and reviewing digital information are important but secondary tasks. Using smartphone technology or its professional-grade equivalent may be an unfamiliar activity for these workers and may be a barrier to its correct use. In a search and rescue operation, volunteers from many organizations and different walks of life are quickly brought together, and their priority is to search, not fiddle around with technology. Ensuring that your fieldworkers are comfortable using your digital tools is the best way to ensure success.

When considering how to train your fieldworkers with new technology, choosing to train in their natural habitat—the field—offers many benefits. As a trainer, you can observe any potential obstacles to the correct use of the technology and help eliminate frustrations and mistakes early on. A positive experience on first use goes a long way toward keeping the technology in hand, not buried at the bottom of a backpack.

The steps that are not even part of the data capture or data validation often impose the greatest barriers. Turning on the device, signing into the app, having the right permissions to view the project, and having an internet connection to see online content or synchronize data are usually the biggest hurdles to overcome.

When the fieldwork team is small, some one-on-one time with everyone working in the field will benefit fieldworkers and trainers. In the case of larger fieldwork teams, the creation of a pyramid training scheme makes sense—not one of those sales pyramid schemes where the person at the top collects all the money, but one where key fieldworkers are trained and then responsible for sharing the knowledge and acting as local support for the rest of their field crew.

A one-page quick guide—with pictures—might be a way to give ongoing moral support to a new user of technology. Showing button combinations to turn the device on (and restart) is great for these paper instructions. Get users to the pages that they are familiar with and out of the menu system of the device as quickly and easily as possible.

When it comes to the actual training, what do you say or show? Ideally, the app, map, or form should be intuitive enough that not much instruction needs to be given. But often there are choices and options that better suit different scenarios, so being on hand early in a deployment to answer questions immediately can help set a good pattern of choice for fieldworkers.

Sometimes training may not be about your app, map, or form, but about the instrumentation to be used and your field device of choice. Anyone who has helped a parent or grandparent with learning how to use a computer mouse will have experienced the need to expand their vocabulary from "click the button" to "click with the right button but only once and don't move the mouse while you click it. Oh no, now you moved it into a folder—which one was it? Let's start again!" A similar challenge of technology instruction is figuring out how to explain how to use a stylus on a tablet or smartphone and the difference between tap, double-tap, and tap and hold. You cannot assume that all your fieldworkers are comfortable with these concepts and can perform the action without extensive practice.

If your fieldworkers will be using a combination of equipment—for example, a smartphone and Bluetooth GNSS receiver—consider breaking up

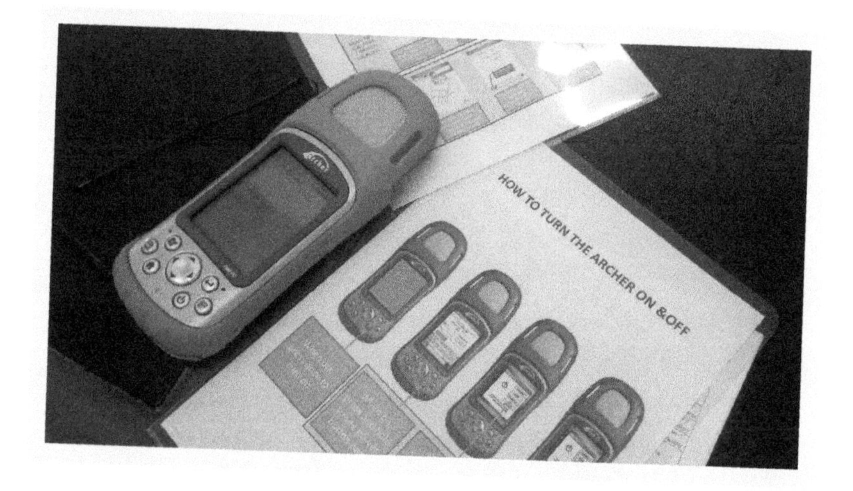

Include a one-page go-to instruction manual for fieldworkers on how to use the device.

the learning tasks, allowing them to get comfortable with your app, map, or form before adding the complexities of Bluetooth connectivity and GNSS correction services. Could you have them use just the internal location sensor in your device of choice to get started? In the "Equipment" section of this book, you can read more about Bluetooth and GNSS receivers and why they are both awesome and a potential headache at the same time.

So, you've stood next to your fieldworker under a tree watching over their shoulder; they have tapped all the correct buttons, and they tell you that they are good to go. Whether they are an experienced user of technology or a novice, be prepared for that phone call from them that says, "It stopped working." What stopped working, the tablet or the GPS? The ability to type and enter a value, or the ability to sync? To take a photo, or to perform a calculation? While you are with the fieldworker, be sure to use consistent words and terminology for the field device and app of choice, so that when they report issues, they use this same language.

In our office, we make a point of calling all the smartphones and tablets on which we install our apps the "device." We also make a point of calling all the GNSS hardware that you connect to those devices "receivers." Just being clear on which is the device and which is the receiver can go a long way toward communicating the situation when you're on the phone with someone who is miles, or even countries, away.

A little trickier for some people might be differentiating between an app and the content of the app that you are using—the map, project, or form. When everything is working well, it doesn't really matter, but when you are trying to troubleshoot, understanding the difference can significantly help isolate the problem and point to a resolution. An issue in the map, project, or form might be quickly fixed by someone in the office, and a refresh of your gallery of projects can have the fix to your fieldworker in moments.

An issue in the app may take a little more troubleshooting. Try another project—does that behave the same? Is it the app, or is it something else in your map, project, or form that is causing this behavior? Can you create a project that is much simpler than your everyday project, and it still fails?

Isolating an issue to clearly show a bug in an app means you can report it for consideration in a future release of the software and, more importantly, work around that bug in the moment and find another way to perform the action you need. The bug may be resolved in a day, a week, a month…or never—but that's mostly out of your control. A healthy description of the issue, why it's critical, and clear steps to reproduce the problem can help bump up the prioritization of a fix. But as apps do more and more, their complexity increases, the to-do list gets longer, and the promise of a fix time is indefinite, so it's always best to have an alternative.

Playing solitaire is work

The game of solitaire has been included with the Windows operating system since 1990, and as well as being a great time filler, it's also great at teaching someone the critical mouse skills needed for everyday operation of a computer. The game was also included on Windows Mobile hardware and worked well to train newbies in the art of using a stylus.

Describing how to tap, double tap, and tap and hold with a stylus on a digital screen can be an eye-watering exercise. But giving someone an app that mimics a behavior that is recognizable—like the action of moving a playing card to a pile—is an excellent way for them to practice this new skill.

Consider adding a game such as solitaire to your bag of teaching tricks if you need to instruct someone in the biomechanics of using a touch screen or mouse-driven device.

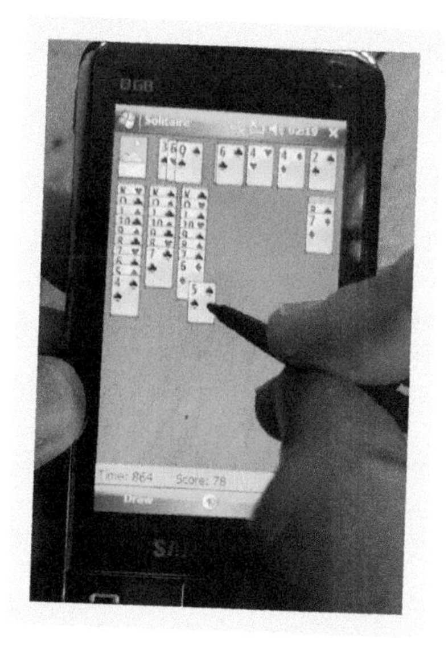

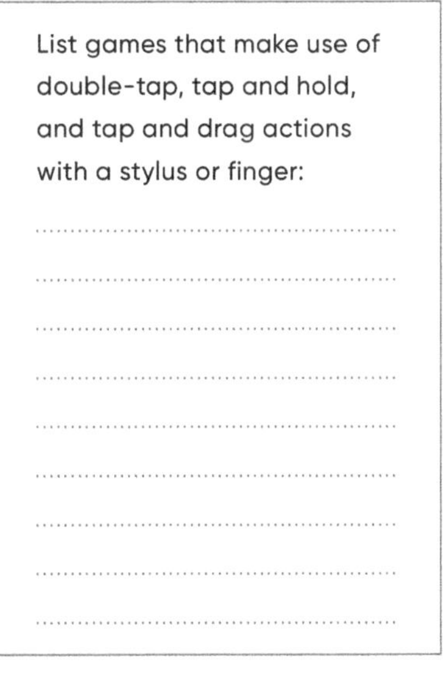

Playing solitaire on a computer can teach basic touch screen skills.

Cascaded training of fieldworkers

Consider this scenario: a gas network maintenance team for a city comprises four field crews of six people each. In each team, some members have been working in the role for many years and have a rich knowledge of the water network and the tools they use to repair breakages, as well as what seems to be a sixth sense about the cause of problems. The team also has some young apprentices. Their knowledge of the network is not as comprehensive as that of their more experienced team members, but they're pretty good at all this technology stuff.

Putting the field GIS in the hands of the apprentices can ensure an early success with capturing data in the system because of their ease with the technology. It can also provide a way for the team to build cooperative learning opportunities and work together, with the apprentices assisting those less familiar with the tech and the seasoned workers showing the apprentices how to do the work.

Creating this two-way learning environment can strengthen the team. The experienced workers share their knowledge about the job, and the apprentices share their knowledge of the technology.

The days of "old dog, new dog" hierarchical teams are gone; teams work and grow best when they can learn from each other. Remember that the biggest hurdle is often turning on the device and signing in. If this problem can be quickly overcome by someone within the team, a tech support call can be averted and unnecessary frustration avoided. This, in turn, leads to a better chance of the form you wanted the fieldworkers to fill in being filled in.

Think about the people on your team or teams, and in the diagram, identify who can be your field GIS champions, your subject matter experts, and those who are somewhere in between.

Training fieldwork teams

Field GIS lead ....

Team 1 ....

Field GIS champion	Mixed-experience workers	Subject matter expert
....	....	....

Team 2 ....

Field GIS champion	Mixed-experience workers	Subject matter expert
....	....	....

Team 3 ....

Field GIS champion	Mixed-experience workers	Subject matter expert
....	....	....

Team 4 ....

Field GIS champion	Mixed-experience workers	Subject matter expert
....	....	....

Allocating tasks

Whether you are working alone, as part of a fieldwork crew, or coordinating many fieldwork crews, creating a methodical routine for your GIS fieldwork is key to completing a successful project. In some cases, subsequent fieldwork choices depend on the outcome of preliminary fieldwork activities. Being able to see the progress of your fieldwork on the fly can allow a team to redeploy resources as needed while in the field.

Consider the example of a preconstruction archaeological survey. The geographic extent of the survey is absolute—100 percent of the designated buffer around the planned infrastructure is to be surveyed. Similar projects in similar locations can provide a good estimate of the time required to complete the transect survey. In this case, assume that the team consists of six team members, and they have 10 days and 190 acres to cover. For this project, the best form of task allocation is to clearly designate a geographic extent for each team member. Working simultaneously on adjacent transects keeps the team close together so that when they take a break, they can discuss progress and adjust where needed during their short study time.

Scaling up the job of allocating tasks dramatically, consider a population census project. Across the world, countries have been conducting population censuses for hundreds (and even thousands) of years. Data capture techniques have significantly evolved, but the principle remains: capture information about the entire population that will help guide research, development, and distribution of resources and facilities to serve that population. In your own lifetime, you may have seen the method of census information capture evolve from filling in a paper form that someone hand-delivered to your door to completing an internet form on your own time.

Moving from paper to digital forms doesn't magically solve the allocation-of-tasks problem—but it does provide new ways to help manage the logistics of a large fieldwork team. Some countries have moved to an online system of census data capture, but even those will have some portion of the data collected by enumerators who go door to door to complete the survey. Enumerators are used extensively in regions where preexisting household data may not be available, literacy is low, or the technology or resources required for citizens to complete the census themselves is simply not available.

Using a quick calculation, to conduct a population census within a two-week window, a country with a population of 10 million people would likely need 10,000 enumerators in the field, at 100 heads per day per enumerator, to capture the census data. That's 10,000 people to train to use your device and app of choice and, even more mind-boggling, 10,000 devices to purchase, set up, and manage. Just take a moment to imagine what 10,000 boxed tablets look like—they would completely fill a 20 ft (6 m) shipping container. Every one of those tablets will need apps installed and projects, maps, and data loaded on them.

Today, there are ways to streamline the deployment of any number of mobile devices for fieldworkers. A mobile device management (MDM) product can be used to manage devices, provide apps, enforce compliance with corporate policies, and support authentication with secure networks. Once each mobile device is identified with the MDM, a preconfigured collection of apps and settings can be wirelessly deployed to that (and all identified) devices. App needs an update? No problem, configure the update in the MDM, and all connected devices will get the update. Deployment by MDM isn't an instant solution; you still need to ensure that each device is compatible, identify each device, and create configurations that suit different makes of hardware. The MDM's mobile app would need to be installed on every device, and an internet connection must be available. But after all that is accomplished, you can deploy with one click.

Sounds simple, right? It never is, but it's a significant advance on individual setup of every device. When you need thousands of devices—say, for the population census described here—working directly with a hardware manufacturer is essential. The manufacturer can significantly improve your deployment and ongoing maintenance experience by deploying key components (such as an MDM app) out of the box.

Now that you have 10,000 enumerators with equipment in hand, allocating tasks must also be methodical and automated. These tasks might be assigned by geographic proximity, by priority, or by size. Apps and scripts that can analyze work progress can reevaluate and adjust task assignments on the fly.

For our population census, accurately defined enumeration areas are critical. In fact, the fieldwork project to update and define these enumeration areas is just as important and complex as the population census fieldwork project itself. Enumeration areas are typically defined to cover a designated number of households that may differ depending on whether the area is deemed rural or urban. The construction of an apartment building—or many—can significantly alter the enumeration area landscape, so it is critical that enumeration areas are redefined before every census.

Choosing apps that can deliver and update task lists to fieldworkers is essential for effectively managing fieldwork teams.

A patchwork quilt of fieldwork

NAME: Jim **ROLE:** GIS specialist
INDUSTRY: Mining exploration **CIRCA:** 2010

In some projects, planning how fieldwork will be completed is more obvious than in others. Once the overall work effort is accounted for, it's time to divide the fieldwork. Individual task allocation can be managed with many different tools: a paper checklist, a spreadsheet, or an app. The grouping of activities may be based on count, time availability, geographic area, or a combination of these.

With even the greatest amount of planning and scheduling, once the team is out in the field, the real world can make a significant impact on these plans.

Some years ago, I spent time in the field on a project that required swaths of scientific data to be captured across thousands of hectares of the Papua New Guinean jungle via helicopter. With careful calculations before going into the field, our team estimated the flying requirements to cover the area at an adequate flight path density, and based on flight times and costs, the fieldwork schedule was planned to be completed over the course of a few weeks.

This was a geophysical survey, with data collected from numerous instruments mounted on the helicopter. To accurately collect the data, the helicopter needed to be flown at a steady and continuous 100 m altitude above ground level. Maintaining 100 m above ground level in mountainous terrain already requires significant flying mastery by a pilot, but to use the data captured on the flight for scientific analysis, the fieldwork team also needed to be sure that the flight paths matched carefully defined survey lines. The resultant mesh of data values could then be used to predict the geophysical characteristics of the environment.

Keeping track of flight paths recorded for a geophysical survey by helicopter.

With all this methodical flight planning and expensive scientific equipment on board, imagine taking off in a helicopter for a day's data collection, heading toward, up, and over a mountain—only to see an unexpected thunderstorm on the other side heading swiftly toward you. This happened to us early in the project, and we had to immediately turn around, the day's planned work in disarray. Not only did the storm impact how many hours of work we could accomplish that day, but it—and every other day's delay and disruption—also created a work planning requirement that was nothing like we'd imagined back in the comfort of the office before the fieldwork started. Our flight path planning soon resembled a patchwork quilt, rather than a neat set of parallel lines. Sections missed from previous flights were covered on subsequent flights, and flight paths on a given day were chosen to suit conditions. The fieldwork time frame increased significantly, from a few weeks to many months, but the work was completed, and new ways to plan and allocate tasks on the fly were created in the field.

Assign project tasks

Knowing how much time and resources to allocate to a project is learned from experience. Keep a record of how your estimations turned out so that you can better estimate next time.

Example project

Project description

Population census of Iceland where 99% of the population lives in urban areas

Data to be collected door to door in 2-week window

Population 376,200 (2022 estimate)

Office-based testing

One record takes 5 min to complete

Complete 84 records in a day

Would take 4,480 days for one person to complete

Would take 320 people to complete in 2 weeks

Pilot testing

One record takes 8 min to complete

Complete 52 records in a day

Would take 7,167 days for one person to complete

Would take 512 people to complete in 2 weeks

Actual project times

One record takes 10 min to complete

Completed 40 records in a day

Would have taken 7,167 days for one person to complete

Needed 617 people to complete in 2 weeks

Project 1

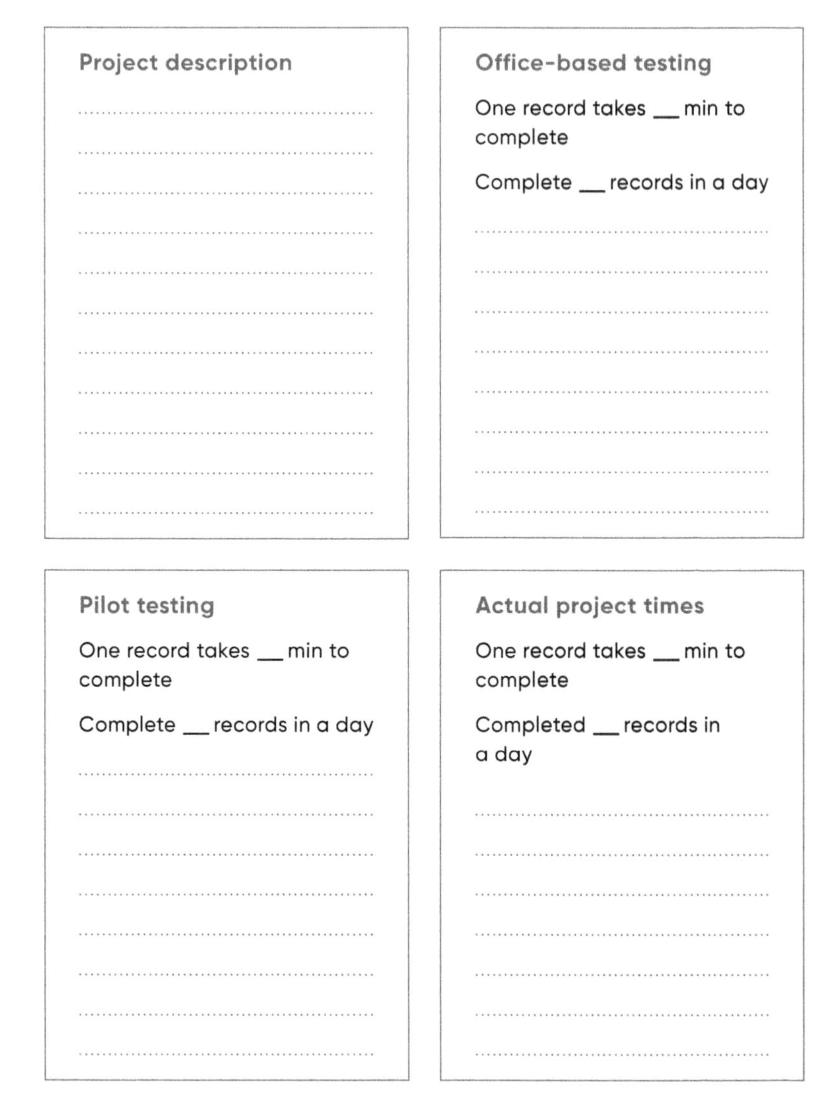
Project description
Office-based testing
One record takes __ min to complete
Complete __ records in a day
Pilot testing
One record takes __ min to complete
Complete __ records in a day
Actual project times
One record takes __ min to complete
Completed __ records in a day

Project 2

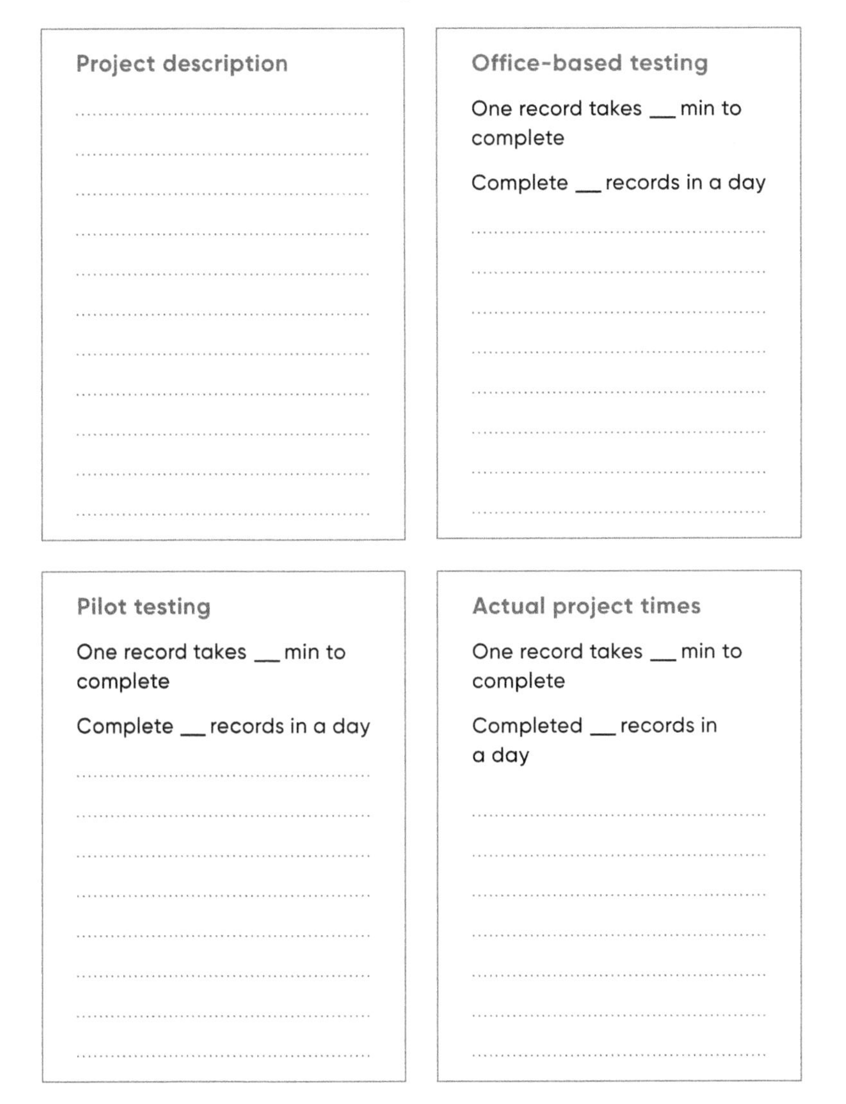

Project description

Office-based testing

One record takes ___ min to complete

Complete ___ records in a day

Pilot testing

One record takes ___ min to complete

Complete ___ records in a day

Actual project times

One record takes ___ min to complete

Completed ___ records in a day

Notes

3

Equipment

Devices

When you're preparing a fieldwork data capture project, it's easy to design and test it in the office on your computer and your smartphone. This is great, for maybe the first one or two iterations only.

But your data collection device and app of choice must be tested by both you and your fieldworkers outside where it will be used, many times over, well before you consider it ready for real work.

Even standing away from your desk outside the front door will reveal significant opportunities for improvement, such as better color choices to use for working outside in bright sunlight. Or you may discover that the device is too heavy to hold while standing (or walking) for long periods, or that the position sensor on the device will return an accuracy of only 10 m at best, even with a clear view of open sky. These first observations of using your field device and app of choice might send you down a rabbit hole of exploration. But that's OK. Better to do that now, instead of when you've deployed a team of people, hardware, and software far away from the office and possibly with no internet connection.

Choosing a mobile device is a completely biased activity. You can create a list of acceptance criteria and test different models of hardware side by side in the field, but ultimately you will pick one. The least expensive, or the most familiar device to you, will most likely be your choice.

In a purely technical comparison when it comes to features, all brands have subtle variations of the same thing. Some people will say (they typically insist) that "iPhones are easier," but really that's because iPhones are what they've used the most (or exclusively). Someone whose personal phone is an Android may say, "Androids are better." They may even be more specific and say, "Samsung is better," not really knowing the difference between the operating system and the hardware.

For those of us who have come from the Palm Pilot and Windows Mobile era of field devices, they are all the same, just with the settings and options in different places. Sometimes things need one click, sometimes they need two. But typically, all brands have a model of device that can do what you need.

The traits of hardware that are far more important than the number of button clicks to the settings are battery life and screen readability in mixed conditions—from standing out in the sunshine to walking into a dark pump house to capture information about a sewer network.

Screen readability can be a showstopper. Modern computers and smartphones make a good attempt at adjusting for lighting conditions. The automatic brightness adjustment on my computer in the early evening sometimes acts as a reminder that maybe I should stop working for the day. Our smartphones can make similar adjustments but are limited by the physical material of their screens. Even some of the best hardware designed for outdoor use can have a tough time in the varied conditions in which you might need to use it.

Even if the screen brightness is OK, you can't use your field device of choice with a dead battery. In those cases, your only field device of choice is a paper notebook. And now is a good time to come to grips with something: a paper notebook is always a good idea for fieldwork. (More about that later in the section.)

For your devices, battery options are vast. Consumer-style devices can be charged by a battery pack that can be as small as the device itself. In turn, these batteries can be charged by 240-volt wall chargers, 12-volt car batteries, or even solar panels. Some commercial devices may require proprietary charging cables and a wall charger, but these units also tend to have hot swappable battery options. Know your power requirements and your recharge options before you consider an app and device of choice.

The duration of use isn't the only indicator to consider when deciding on battery choices. Consider other aspects of the hardware. Will you be carrying the device in your hand all day? Will you be taking it out of your backpack every half hour and holding it for just a few minutes? Do you have a risk of dropping it on a rocky surface, or are you working close to water?

Think about the mechanics of using your app and device of choice in the field. Imagine it's winter, and you're wearing gloves. You're not a fan of smartphones—your kids do your online shopping for you. You're six feet tall, with correspondingly large hands. You're carrying a backpack with a few glass jars to collect samples, a pH meter, a paper notebook, and the app and device of choice that was given to you as you rushed out the office door. You sit down on a rock ready to take a water sample and open your backpack. How excited do you think you'd be to see that little computer staring back at you from the bag? I know I'd be grabbing for the notebook and saying to myself, "I'll do it later." And probably forget to do it for a few days. And then not be able to read my own handwriting and realize I forgot to write down the time that I collected the sample.

If you're nodding and smiling in recognition, you already know the benefits of digital field data collection, and it's exactly this scenario ("I'll do it later") that you need to plan against when choosing a device.

The all-in-one dream

NAME: Marika ROLE: GIS specialist
INDUSTRY: GIS consulting CIRCA: 2009

As an early adopter of technology, I have drawers full of the "next best handheld computer." No matter how good it looks on the box, any device that claims to be an all-in-one always has its drawbacks. Be aware: one of the critical issues that these devices are trying to eliminate (having to manage and carry multiple pieces of hardware) is, in fact, a key advantage—having multiple pieces means you can choose the best of everything.

In a mood of nostalgia, I recently pulled out of one such drawer a Samsung Omnia i900. On its release in 2009, this was the ultimate all-in-one device, "perfect" for fieldwork. With front and rear camera, Bluetooth, Wi-Fi, a GPS, a one-of-a-kind finger touch mouse, plenty of apps to download, and a size that could comfortably fit in a shirt pocket, what more did you need?

Well, the camera worked great—if you were just taking pictures. And the GPS worked great—if you were just capturing features on a map (and you were in open space, in town, with cell phone reception). And the apps worked great—if you didn't try to use them all at once. Attempting to use all these cool features at the same time ground things to a halt. Oh, and I forgot to mention, the screen readability in bright sunlight was awful.

Why is this story about an old smartphone relevant now? Because, even today, we are constantly being sold the next game-changing all-in-one device. The changes now are more subtle and might be harder to comprehend (what really is the practical difference between one microprocessor or another?), but if you get caught up in the marketing hype, you will just have to have the hot new device.

Having said all that, even after 14 years, I still find the finger touch mouse of the Omnia awesome. I don't know why those didn't take off!

Choose a device

Use the following four flowcharts to choose battery, data, screen, and location measurement requirements for your fieldwork project.

Choose a device based on batteries and power

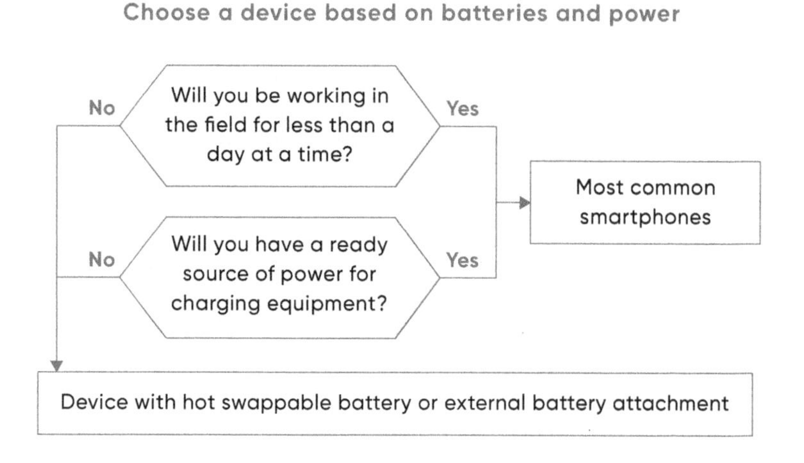

Choose a device based on data and memory

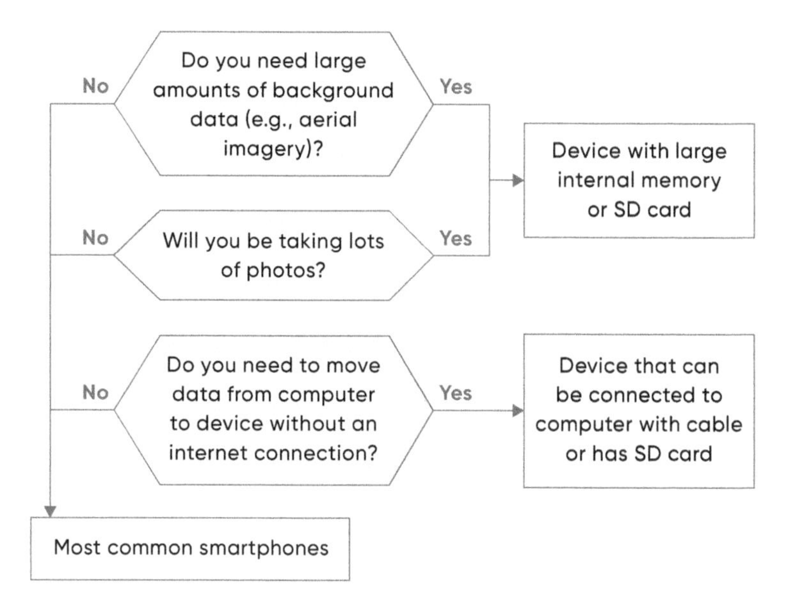

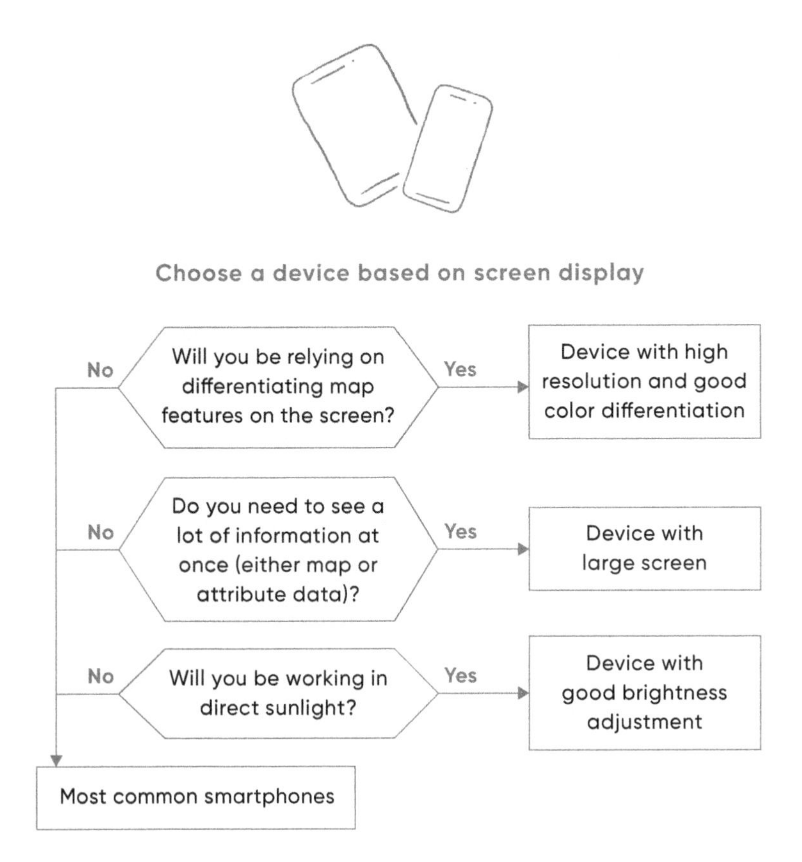

Choose a device based on location measurement

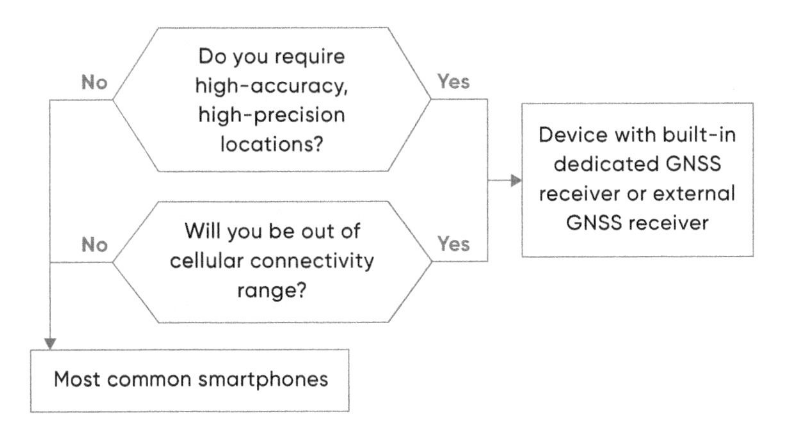

Device requirements

Circle the best option in each box to help describe the characteristics of the device that best suits your project.

Battery / power

Most common smartphones

Device with hot swappable battery or external battery attachment

Data / memory

Most common smartphones

Device with large internal memory or SD card

Device that can be connected to computer with cable or has SD card

Screen display

Most common smartphones

Device with high resolution and good color differentiation

Device with large screen

Device with good brightness adjustment

Location measurement

Most common smartphones

Device with built-in dedicated GNSS receiver or external GNSS receiver

Notes

Which app is the best?

There are so many apps to choose from—which is the best?

When network connectivity is a given, entering information on a web page that is directly connected to your organization's GIS may present the lowest level of complication for a field-to-office workflow. But if you're reading this book, it's likely you have a few more criteria that you must consider when choosing your app.

In the old days, we called them applications. You used disks to install them on a computer and then transfer them to a device through the computer. Some applications did a lot of things (most did only one thing well), and some were customizable. Fast-forward to the 2020s: apps are now downloaded wirelessly within seconds, some still can do a lot of things (many still only do one thing well), and some are customizable.

Doesn't sound much different, right? What's new is the sheer volume of options to choose from. So, which one is right for your project? There is no single answer, but some key traits are worth looking for when selecting an app for your project. Typically, there will be one or two apps that suit most of your projects, and you can use them over and over. But don't be scared to branch out on a project because you need to do something differently. You may be able to customize an app that you already use to suit the new project, but it may also be that choosing a completely different one is a better option.

What kind of data is your fieldwork focused on? Are you capturing or reviewing locations on a map, or are you filling in or editing a lengthy data entry form? Maybe some of both? Do you have the time to stop and enter detailed information at each location, or do you need to be able to capture data while moving, either walking or as a passenger in a motor vehicle or aircraft? If multiple apps can work with the same underlying data, choosing a different app for each project within a suite of projects is a great way to ensure that you always use the best tool for the job.

Are you going to be using a device that is always connected to the internet? If this is an easy yes, perhaps an app that runs on the web is right for you. The term *web app* is a little odd. Isn't it just a page on the web? Not really. A web page has information on it that you read; there are buttons that allow you to jump around to other parts of the website, and perhaps a few text boxes for you to submit feedback. Its primary objective is to give you information. A web app will let you do a whole lot more. Various buttons and switches and options allow you to change how the information on the page is shown. Web apps encourage interactivity with the information and prompt you to add your own.

A web app for fieldwork is a good choice when you need to fill in some values and submit them quickly to the central database or to look up the latest available information for your geographic location. You usually access a web app by browsing to a web address, and you can pin these pages to your device's home screen and return to them as often as you need. Nothing is installed on your device, and data is always current and live.

An app that you download to your device—from an app store or through your computer—might also use the internet to work. But these apps are usually designed to work offline, and you choose when to perform online actions such as synchronizing data that is critical for fieldwork. In some cases, such actions might be continuous; in other cases, it might be hourly, daily, or (most risky but sometimes inevitable) weeks after completion of the fieldwork when you return to the office.

If offline usage is the primary characteristic of apps that run natively on your device, their ability to use hardware onboard the device, and hardware connected to it, is a close second.

Most devices have an onboard location sensor, but connecting a Bluetooth GNSS receiver to a device is a common practice in fieldwork and requires use of a native app. Other sensors directly on the device can also be used by native apps. The accelerometer, compass, and tilt sensors can all communicate to a native app about how you are holding or moving the device and provide information that can inform your fieldwork—for example, the speed and direction of your travel.

Native apps are built for the operating system they run on. Today's fieldwork is typically performed on one of three operating systems: Windows, iOS, or Android. The prevalence of each is strongly skewed by industry and geography. Particularly with iOS and Android, the consumer device market influences the availability of hardware in a country.

Using what's available to you is an incentive you cannot dismiss. You may think that Windows is just a desktop operating system, but rugged (and not so rugged) laptops and tablets have been used in fieldwork for years and are still often the device of choice when the device is mounted on the dashboard of a truck or used as a primary device at base camp. When handheld portability isn't a requirement, why wouldn't you choose the most powerful mobile computer you can get? This in turn means that your operating system of choice may dictate the native apps that you can use. It's reasonable to expect that most apps you need are available for your operating system, but it's wise to check before you invest time and effort.

Choose an app

To choose an app, you must consider your project requirements and the functionality that different apps have. A good place to start is to decide whether a web app or native app suits your needs. This may significantly trim the list of apps you need to evaluate.

Next, you should compile a list of activities you need to do or functionality you need to use. Knowing what you need will make it easier to choose an app.

Web apps versus native apps

The following table lists some of the most common pros and cons of web and native apps. Circle the features that are important to your project to help you choose which type of app is best for you.

	Web app	**Native app**
Pros	Allows direct editing of information contained in connected databases. Doesn't require installing an app on the device. Web browser version dependent but not operating system dependent. Not installed on the device, but can be launched by clicking a button on the home screen.	Typically designed to work offline. Data can be captured and synchronized later. Can use sensors on the device (location, compass, accelerometer, camera). Can use connected equipment, typically Bluetooth location sensors. Can use operating system security features.
Cons	Must be connected to the internet (some caching of single responses possible).	Requires app installation files to exist and be deployed for each operating system.

Types of fieldwork

Think about the type of fieldwork you do, and create a list of the activities you do or functionality you use most of the time. This requirement list will point you to the web or native app that you can configure for frequent use.

Activities I do often:

....

....

....

....

Functionality I use often:

....

....

....

....

The app that does this the best for me is:

....

One-off projects

What about one-off projects? Projects that may be bigger than normal or have different data capture requirements? There's no need to compromise on functionality because it's out of the ordinary. Instead, it's a great opportunity to try new tools and technology.

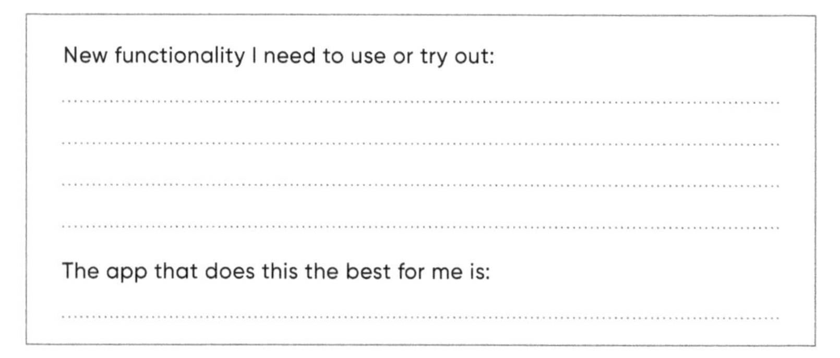

New functionality I need to use or try out:

....

....

....

....

The app that does this the best for me is:

....

Notes

Paper notebooks

The scientific field notebook has formed the basis of ecology, biology, behavior, and conservation studies for hundreds of years. Libraries are filled with cataloged notebooks, and many have made popular best-seller lists. Field notebooks have a life far beyond the scribbled reminders made at a single point in time to remind the author where to start on the next day of fieldwork.

This fieldwork handbook itself was inspired by a book about field notebooks. *Field Notes on Science and Nature*[*] is a beautiful anthology of fieldwork and record taking. Editor Michael R. Canfield has brought together stories from field scientists and naturalists that demonstrate techniques and experiences that can be adopted in many types of field projects. The varied perspectives in *Field Notes on Science and Nature* raise questions about what may be gained or lost with the implementation of digital notes. Its authors share their various opinions on photo taking, database saving, information transcription, and the hazards and blessings of multiple forms of data capture.

In the years since the book's publication, digital tooling has come a long way for many fieldworkers. Your field device and app of choice can make digital data capture a breeze, and a much wider range of connectivity means that syncing and backup of data is almost seamless. Still, there is inevitably a need to "make a note" in almost everything you do. Even with myriad computer screens and tablets at your work desk, there is still a place for a scribble pad for quick notes and reminders—for formulating an idea or for sketching a diagram while someone is describing something. In the field, you may note something that you need to bring on the next day's expedition or a new research question that presented itself as you were performing your current task. You may draw a sketch of an unusual find.

[*] Michael R. Canfield, ed. 2011. *Field Notes on Science and Nature*. London: Harvard University Press.

Note fields in a database table can help. Providing a place for free-form text can be a great way to capture unplanned or unexpected descriptive information. But if there is repeated capture of similar notes, you need to ask, "Why is there not a field specific to this information in the database?"

Rapid expansions in technology mean that you may have not only free-form text in your digital form but also a place to sketch or draw. But keeping up with technology can be both a steep learning curve and a costly exercise, so as your fieldwork evolves, consider the multiple benefits of a hybrid solution. Photographing an illustration from a paper notebook into your digital data collection form is a good way to capture all the fine detail a pencil can offer, creates an instant backup of your paper record that is linked to its digital counterpart, and leaves you with an easily scannable resource—your paper notebook—while working in the field.

There is no question that e-book readers, tablets, and smartphones have revolutionized book reading, but the two features of a physical book that readers miss most are seeing the cover every time they pick up a book and being able to flip through it quickly. For a field notebook, seeing the cover is not critical, but flipping through the notebook is an inherent part of project work. What did you leave out yesterday? What was the color of that same flower you saw last week? Is this one bigger or smaller than the one you saw some other time? Searching by record ID or date or name in a database is good if you know what you're searching for. But flipping through a book and using memories to refine your search is irreplaceable. As you flip through, other information can surface that provides clues to something you didn't even realize you were looking for.

Even with the latest technology, a paper notebook is always a good idea for fieldwork.

Maria Sibylla Merian

Maria Sibylla Merian, a naturalist whose interests focused on the metamorphosis of insects, was the first European woman to independently go on scientific expedition in South America (1699), predating Alexander von Humboldt's expedition by 100 years. She first published her sketches and notes, *Metamorphosis Insectorum Surinamensium*, in 1705, and since then her illustrations and paintings have been used to illustrate many books and as decorative paintings and have played an instrumental role in our understanding of the life cycle of insects.

DE XLIV. AFBEELDING.

ROcu is een groote Boom, brengt roodachtige bloeisel, gelyk in Europa de Appelboomen, als de bloeisel afgevallen is, komt een zaad-huis, dat langwerpig rond, en stekelachtig is als de Castaniën, daar in leggen zeer schoone roode zaaden, deze leggen de Indianen in water te weiken, dan weekt de roode verf daar af, en zakt op de grond, daar na gieten sy het water allenskens af, en droogen de verf, het geen op de grond legt, daar sy alderlei figuuren op haare naakte huid mede schilderen, het geen haar cieraat is.

Onder op de steel kruipende bruine Rupse, met geele streepen en roode hairen, eet deeze groene bladen: den vierden April is sy my verandert en tot een hard en hairig Poppetjen geworden, uit het welke den 6. May zulke donker groene Uilkens voortquamen.

Nach vond ik op dezen Boom bruine Rupsen, als boven op het blad eenen legt, nuttigde deze bladen; den 26. Maart zyn sy ingesponnen en tot een Poppetjen geworden, als een tusschen de bladen legt: den 10. April quam daar uit een zulke grauwe Uil, gelyk boven een zittend vertoond word.

Deeze Boom is de Urucu *by de Hr:* Piso *beschreven, en onder de naam van* Orleana vel orellana folliculis Lappaceis Hermani *werd sy ook in het eerste Deel van den Amsterdamsche Hof beschreeven, alwaar noch andere benaamingen van deeze Boom gevonden worden. D'Heer* Tournefort *heeft deze Boom, als een nieuw Geslacht nevens de twee soorten van* Cortusa Americana, *voor gestelt onder de naam van* Mitella, *dewyl de vrucht van deeze Boom, gelyk als die van de twee soorten van* Cortusa Americana, *ryp zynde, open splyten, en alsdan een kleine Myter ofte Bisschops Muts verbeelden, en noemt dierhalve deeze Boom in syn Institutiones rei herbaria* Mitella Americana, Maxima, Tinctoria.

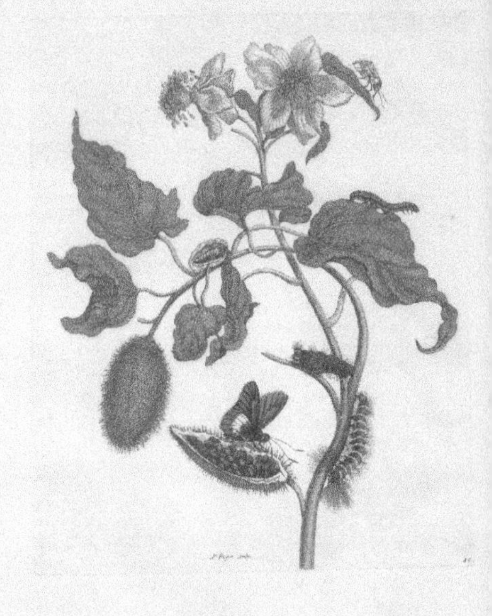

St Louis October 8th 1806

hereas we the undersigned having on the 19th Oct.
gaged John Shields to accompany us on a Voyage of Disco
ough the Continent of North America to the Pacific Oce
l then in behalf of the United States bind ourselves
ow the said John Shields for his services on that Exp
compensation in Lands equal to that granted by the
ates to a Soldier of the Revolutionary Army —

Now Know ye that the said John Shiel
ving faithfully complied with the several Stipulati
his engagement, the undersigned in their said capa
hold themselves bound to the said Shields his H
assigns for the quantity of Lands above Stipulate

Given under our hands the day and da
ove mentioned —

Lewis and Clark

Meriwether Lewis and William Clark's expedition journals, containing thousands of pages of text, illustrations, and hand-drawn maps, are a comprehensive archive of a journey from Illinois to the Pacific Ocean. These journals are now available online at https://lewisandclarkjournals.unl.edu for all to explore.

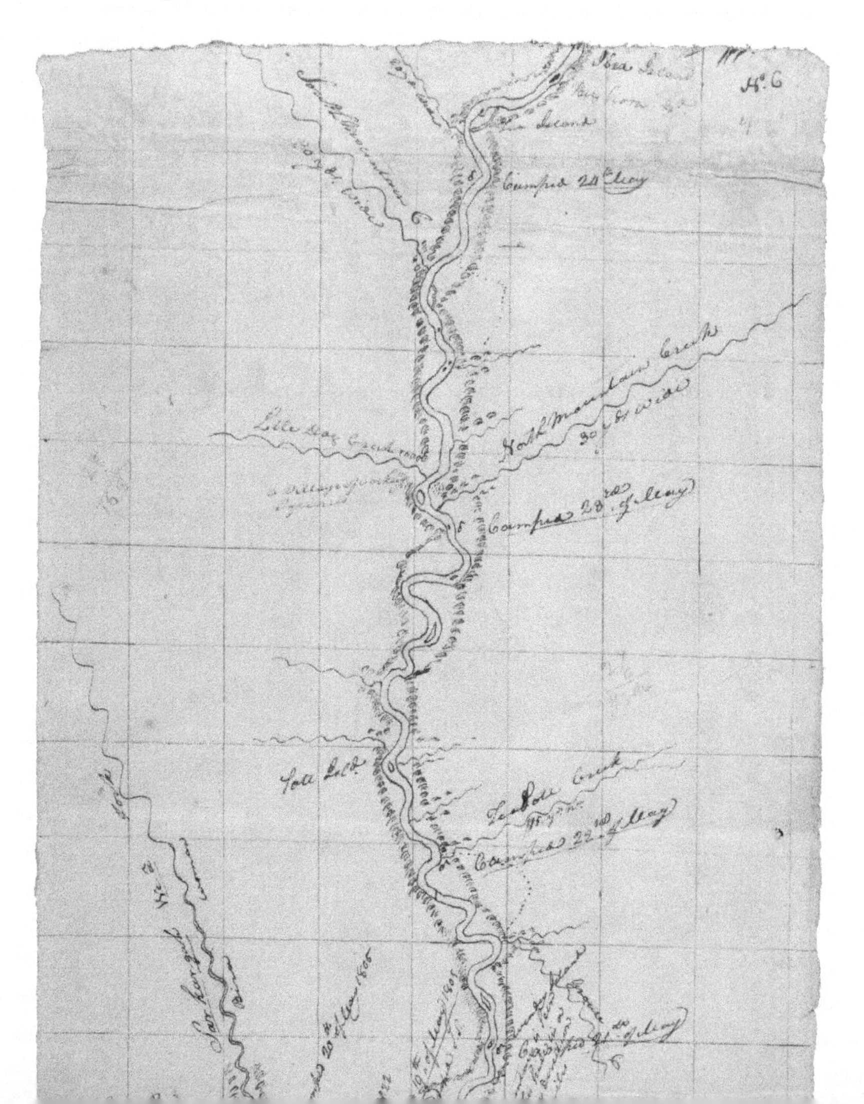

Charles Darwin

Charles Darwin's *The Voyage of the Beagle* (this book and all of Darwin's publications are available at http://darwin-online.org.uk) stands on the shelves of popular bookstores to this day. This book remains in the public eye because it shows Darwin's new ideas on evolution coming to life. Originally written and published in 1839 as one of three volumes (the other volumes edited by Captain Robert Fitzroy) that documented the ship's journey, Darwin's notes are part travel journal and part scientific observation. His writing style was popular and prompted republication of his volume on its own. In the second edition (1845), Darwin expanded his original thoughts on what would later become his natural selection theory, and hence the book has stayed in the sights of the scientific curious to this day.

1. Geospiza magnirostris. 2. Geospiza fortis.
3. Geospiza parvula. 4. Certhidea olivasea.

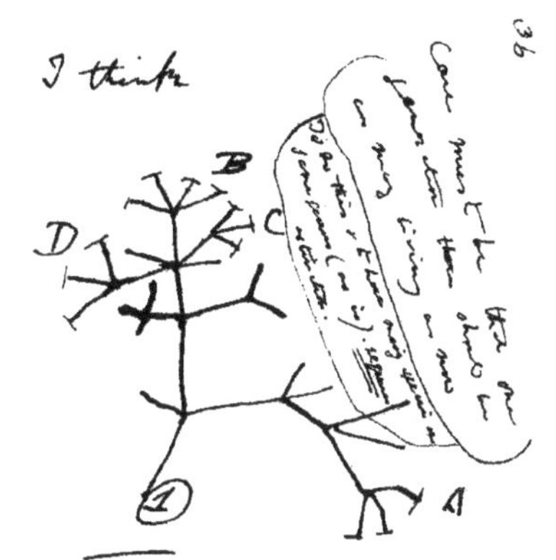

36
I think
1
A
B
C
D
Then between A & B. immense
gap of relation. C & B. the
finest gradation, B & D
rather greater distinction
Thus genera would be
formed. — bearing relation

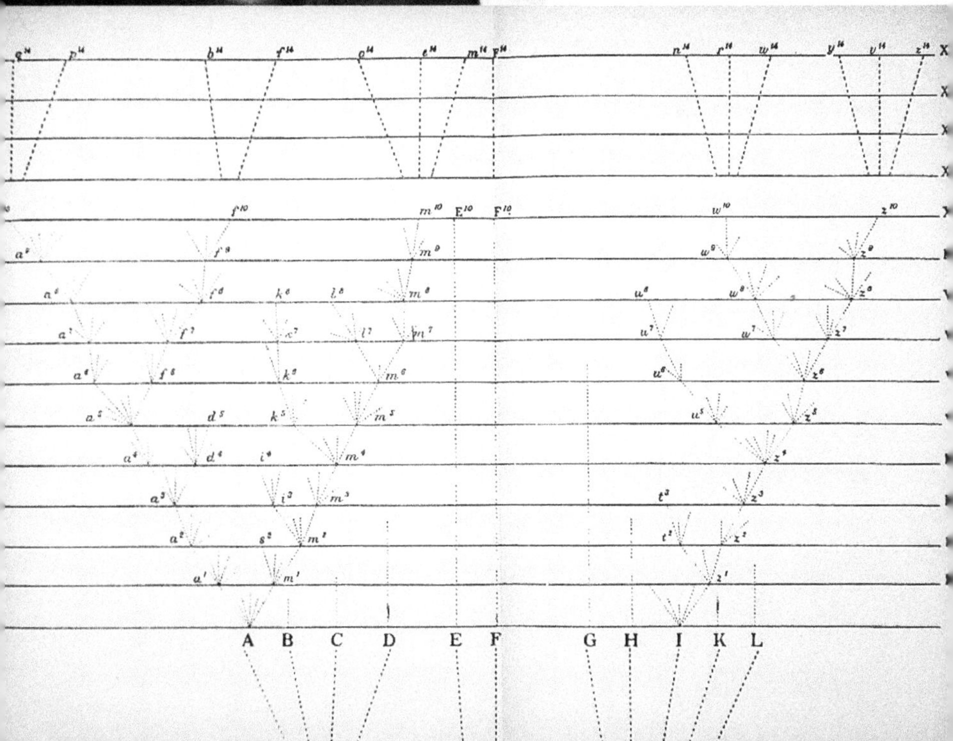

A B C D E F G H I K L

Notes, ideas, inspiration

GNSS receivers

The most common piece of hardware (along with a tablet or smartphone) that a fieldworker will carry is a Global Navigation Satellite System, or GNSS, receiver. Most modern mobile devices have an integrated location sensor, but in general, its accuracy should only be considered OK. Step out of cell phone coverage, stand among high-density buildings or in a forest, and your confidence in location accuracy will quickly evaporate.

The steady stream of data sentences on location that are sent from geolocation satellites are best received and interpreted by a dedicated GNSS receiver. These data sentences contain everything needed to calculate a location with quantifiable accuracy. A receiver that can use all this data and immediately postprocess it along with available ground-based location information requires specialized hardware and software. GNSS radio technology can also be power intensive, so mobile phones don't always have full GNSS hardware on board. Instead, a simpler component may be used that is assisted by other phone components that can contribute to the location determination. By triangulating signal strengths within another well-known ground-based network—the cell towers from which a phone's communication signal is obtained—you can derive the location of the phone. Ideal for using a location-based app in the city, but not great for capturing the location of an invasive ground-covering plant in a dense forest.

Using a dedicated GNSS receiver can not only offer much more detailed information about your location but also help you distribute your computing and power requirements. Consider the record-keeping work of a park ranger on patrol: capturing records of natural hazards, identifying trash or damaged park facilities that need to be attended to by others, or updating publicly accessible information about the park. Using a consumer-style mobile phone device to take photos, capture locations under a dense tree canopy, and enter descriptive information about a location can quickly chew through battery power that was originally designed for a few phone calls and casual internet browsing over Wi-Fi. Alternatively, pairing the mobile phone with an external GNSS receiver can conserve battery power on the mobile phone and provide a better location for the records collected.

Of course, you must consider how your receiver will be connected to your data capture device. The most robust connection may be a cable connection, but the most convenient may be Bluetooth. When considered in isolation, neither connection type is better than the other. How you are going to carry your equipment, how long you will use it, and what level of location accuracies you require must all be considered when choosing a GNSS receiver.

GNSS receiver log

Keep a log of the GNSS receivers you have used, tested, or evaluated and save for future reference.

Make	Model	Compatible OS			Proprietary app required? Name
		iOS	Android	Windows	

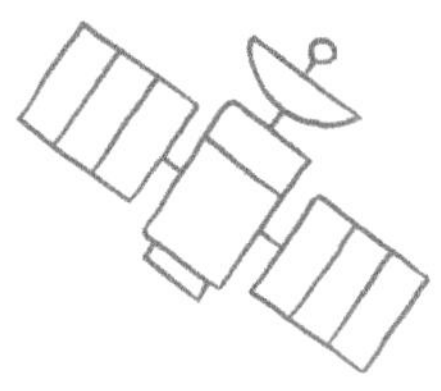

Comments

Accuracy, battery life, project suitability

Bluetooth connectivity

Bluetooth is a short-range radio connection originally created in the 1990s, and still primarily used today, for wireless headphones. The communication protocol is packet based, which means it's also ideal for transmitting location data in the form of a comma-separated data sentence from a GNSS receiver to your device and app of choice. Once a connection is established, little will go wrong with transmission. Those sentences will be passed along as a stream of messages, your device and app of choice will translate those messages into position information, and the blue dot on your map will show you where you are.

When working well, Bluetooth is a fieldwork game changer. A GNSS receiver can connect to your device and app of choice and provide accurate and repeated location measurements. Just like those awesome headphones you bought to use with your new phone. But did you ever try connecting them to your computer to see if they were better than the built-in speaker for meetings? And now do they mysteriously sometimes connect to your computer when you're really trying to connect to your phone? Or do they connect to your car (because just once you thought you might try them out in the car)?

Bluetooth is a fickle friend. Just when you think you've got your connections sorted out, it will laugh in your face. On that day when your Bluetooth connection is just not connecting, think back to reading this and know that the only thing you can do is turn everything off and on again (and maybe repeat several times).

Most people using Bluetooth in fieldwork will typically have one device and app of choice and one GNSS receiver. If your field team has several GNSS receivers, it's best to assign them to individuals, or, better yet, assign them to a buddy device and app of choice. Connecting to the same receiver over and over will present no issues, or very few. Do this and keep your fieldwork (and sanity) running smoothly.

The most common issue when working with an external GNSS receiver is the connection between the receiver and your device of choice. If the receiver is not listed in your app of choice—or is not displaying any location information—go back to your device settings and check the connection.

Very few GNSS receivers will allow connection to multiple devices; typically, it's a one-to-one connection. But if you were using another device this morning or yesterday, or someone else borrowed the receiver to try it out, there's a good chance that another device is clinging to your receiver, stopping any new connections. Just like your car and your headphones. If it really was an experiment, and you're not going to use the headphones in the car again, unpair them. If you are testing your GNSS receiver with different devices, be sure to disable Bluetooth on all other devices while you test the receiver you're interested in.

If your pain persists—or, more accurately, your Bluetooth connection does not persist—it is still more likely a hardware issue. More specifically, firmware and operating system compatibility.

Every piece of digital hardware has its own firmware. Go back to the manufacturer's app and ensure it is up-to-date. With our ever-updating world of operating systems, hardware manufacturers must keep updating, too. Getting your receiver-to-device connection rock solid will give you by far the best chance of success with your app of choice.

Bluetooth blues

NAME: Brett ROLE: GIS specialist
INDUSTRY: Forestry CIRCA: 2018

Commercial forestry companies manage native and plantation forests for their environmental, social, and economic value. An accurate knowledge of tree type, age, size, and count can maximize yield and quality of products while also allowing the company to be a good neighbor, doing the best for the environment.

Our team of foresters appreciated that we could collate more accurate information about the forest areas, or coupes, that we managed, both the existing native hardwood forests and our existing and future plantation forests. We were keen to make use of GNSS technology to better measure the size of each coupe managed by the company.

The team members were also conscious of choosing the right tools for the job so that, going forward, we could capture with minimum effort the boundary of new plantations as they were set up. For now, we planned to capture boundary areas for all the existing coupes. Some of these were up to 30 years old, and many had been hand drawn on paper maps. Of course, the boundaries hadn't moved on the ground, but making a better measurement of the boundary now would mean better harvest planning in the future.

The equipment requirements were minimal: a smartphone, an off-the-shelf app, and a Bluetooth GNSS receiver. No messy cables—the Bluetooth connection was all we needed. Because there was limited internet connectivity in the forest, the locations would be postprocessed to improve accuracy after the data was captured. So, if the data could be saved on the device, the foresters would have few complications in the field.

As the GIS manager, I planned a pilot project and sourced several Bluetooth receivers on loan so the team could test them in the field and make a thorough assessment before we invested heavily in the equipment.

Approximately 10 foresters would be using this technology during the pilot, but when it was time to cover all our ground, we would bring in an additional 30 contractors to assist.

I had a good understanding of the technology and had already earmarked a particular model of GNSS receiver that suited our needs. It would provide the required level of precision, was compatible with our mobile devices, and presented the fewest complications for use in the field. This was going to be my recommendation; I just needed the team to confirm they were able to use the receivers.

I knew that most of our foresters had a reasonably good relationship with technology, but I still prepared as much instruction and support as possible. I did demonstrations, created cheat sheets, and simplified the process to a minimum. But their most repeated complaint was connecting the receiver to the device using Bluetooth. We talked through it over the phone, and I continued to test with the same hardware combinations in the office. On one occasion, I traveled several hundred kilometers to one of the plantation sites with one of the foresters to see if I could spot an error in his methodology. There wasn't one. It just wouldn't connect. We reset the receiver, we reset the smartphone device. We tried over and over. We were doing everything right but still no connection.

After a lot of frustration, the group of foresters found that a different-model receiver offered the most reliability in Bluetooth connection. They were able to connect with ease and move on to the mapping tasks that formed the key part of the trial. The receivers worked well, and the foresters were keen to move forward with the project using those receivers—which cost 10 times more than the model I'd planned to recommend to the decision-makers.

Considering the cost, I was wary of the response to the outcome of the pilot. But the sanity (and human labor cost) of the foresters prevailed. The more expensive units were purchased, and the coupe mapping project went ahead with success.

The different-model receivers came from the same manufacturer, the

Bluetooth technology was the same in both, and they were being used in the same outdoor environment well away from possible radio interference. There was no logical reason why the connection was different between them. But there clearly was a difference, and the pilot was invaluable in identifying this for the sake of the fieldwork.

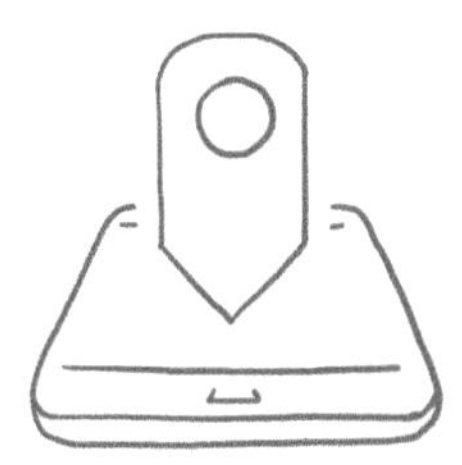

Notes

Bluetooth connection troubleshooting

When working well, Bluetooth connectivity feels like magic: physically disconnected hardware units communicate with each other as if they were one device. When Bluetooth is not working, user frustration can escalate quickly. There's no cable to pull out and put in again, and there is typically little to see on the device. In most cases, it is the connectivity between the external hardware and the device that is the problem. Once the device is paired with the external hardware, apps can see it. The following outline shows a troubleshooting path when working to connect a GNSS receiver to a computing device using Bluetooth.

Basic troubleshooting

Most problems can be fixed with these basic troubleshooting techniques. Spend most of your time and effort here. Don't be afraid to restart often, particularly in the troubleshooting steps for connection and app issues.

- **Restarting**
  - Restart the receiver.
  - Restart the device.
  - On the device, turn Bluetooth off and on.
- **Environmental factors**
  - Ensure you are using the receiver in its optimal operating environment.

    For instance, the receiver might need a clear view of the sky.
- **Hardware issues**
  - Verify that your device can connect to other Bluetooth receivers.
  - Verify that your receiver appears on the list of Bluetooth receivers on at least one device.

    If your receiver appears on one device and not another, then the issue is with that device. If the receiver doesn't appear on any devices, the receiver might not be set up properly or might be faulty.

Connection issues

Under Settings > Bluetooth, attempt to pair the receiver.

- **Receiver doesn't appear on the list**
  - Try "basic troubleshooting."
- **Receiver appears but doesn't connect**
  - Turn off Bluetooth on nearby devices, particularly those that might have been connected to the receiver in the past.

    Most receivers can only connect to one device at a time, and some devices (particularly iOS devices) "hold on" to a pairing.

App issues

Under Settings > Bluetooth, it shows that the receiver and device are paired, but the app isn't connecting.

  - On the Bluetooth menu, try unpairing the device and pairing it again.
  - Verify that no other apps on your device are open and connected to the receiver.
  - Ensure that your app(s) are up-to-date.
- **Connecting to the receiver's proprietary app**
  - If your receiver comes with a proprietary app, ensure that you can connect to this app before attempting to connect to third-party apps.
  - If your device uses Android and mock location sharing is required, enable it in the app (or in Settings).

    This is not always needed, but now is the time to configure it; check your receiver and app requirements.
- **Connecting third-party apps**
  - Verify that the third-party app is compatible with your receiver.
  - Try running the third-party app with the proprietary app running and connected in the background.
  - Try running the third-party app with the proprietary app disconnected and completely closed.

Memory

Knowing what kind of data you are going to use and capture before you go into the field can help you make better equipment choices.

If you know you are going to need large swaths of aerial imagery that doesn't get updated very often (or at all) during the lifetime of your project—for instance, for a geologic survey project—you may choose to copy those image files to a memory card in the office and slot it into the device (or many cards for many devices), minimizing the need to synchronize data over the internet.

Alternatively, background layers for your project may be coming from another team—perhaps an advance party is clearing forest trails, updating and capturing access routes, and you are following behind, conducting a fire hazard assessment. You may need plenty of memory onboard your device, and perhaps all the data, background and operational, needs to be synced every day. In this case, plenty of internal memory and a solid data connection would be critical.

For a condition assessment, the most important type of record is photos—and lots of them. Memory cards can be good for storing photos; your app of choice may have an option to save a scaled-down image with your data inside the database, but you may also choose to keep the original high-resolution version of the photos for specific research. Whichever way you store your photos, knowing which photo corresponds to which record is critical. Where was the photo taken and when, and what is it a picture of? If you know in advance how the photos will be used and stored long term, you can organize your file and folder names from the start to make that job significantly easier.

In field environments where a data connection is costly or not available, planning for multiple ways to back up data is important. Whether it's a three-day or three-week visit to a remote study site, it is important to avoid the temptation of assuming that your one laptop with its offline apps and plenty of memory will be fine. With external memory options now very small (both in cost and physical size), it's essential to build into your fieldwork a routine of backing up data to a USB stick or memory card. These tiny disks and drives will be priceless when the computer goes overboard, gets dropped out of the truck, or stamped on by a frightened animal.

Memory requirement estimates

The two biggest memory consumers are offline basemaps and captured photos. If you head into a forest with little or no network connection and field data cannot be uploaded to a central repository, you could be requiring or saving a lot of valuable information on your device for lengthy periods of time. The following scenarios describe such circumstances and show how you can easily have gigabytes of memory needs. Use the given examples to estimate your own project memory requirements in the boxes provided.

Consider the following questions when estimating your own project memory requirements:

- How regularly can you check in, sync, or upload collected data?
- Do you have the network bandwidth to transmit the data you need to send?
- Do you need to consider doing backups in the field (USB sticks, memory cards, external hard drives)?

Example project

How much background data (basemaps) to be carried?

Scenario

Consider a freight rail construction project between two major cities, spanning 1,600 km across a predominantly rural landscape. Internet coverage is restricted to small areas near townships, but most of the corridor has no internet coverage. An environmental audit team works alongside construction to ensure prescribed land management requirements are adhered to. Aerial imagery tiles need to be stored on a device for use when offline to compare site visits to preconstruction conditions. Imagery for a 1 km buffer around the rail corridor is required.

Assumptions

- Geographic features that are equivalent in size to midsized buildings need to be identifiable, so level 20 map tiles are required.
- Level 20 map tiles are approximately 0.00025° of longitude wide, and 1° of longitude is approximately 50 km. *See "Zoom Levels" table.
- 3,200 sq km of tile coverage is required.
- 1 tile is 0.1MB in memory (0.0001GB).

How much?

Approximating the 3,200 sq km area as a square area (for simplified calculations) results in a longitudinal length of approximately 56 km, which is approximately 1.12° of longitude.

Level 20 tiles are 0.00025° wide; therefore, 4,480 tiles are needed to cover a width of 1°. Approximately 4,480 × 4,480 tiles would need to cover an area of 3,200 sq km = **20,070,400 tiles**.

20,070,400 tiles at 0.0001 GB per tile is **2,007 GB** (just for level 20!).

Example project

How many photos to be captured?

Scenario

Consider a liquefied petroleum gas (LPG) fueling station safety audit project. At each audit site, four photos are taken of the LPG storage unit (for example, north-, east-, south-, and west-facing photos), at a wide enough view to include adjacent buildings and structures and high enough resolution to identify hazards. Images must be saved for seven years for audit purposes, with the previous year's images available for viewing at the time of audit for comparison. According to the Alternative Fuels Data Center: Propane Fueling Station Locations (energy.gov), as of June 2023 in California, there are 15,841 stations.

Assumptions

- All stations are audited annually.
- 2 MB photos are captured and saved to ensure sufficient quality and image size for review by auditors.

How many?

2 (MB) x 4 (count) x 15,841 (stations) = **129 GB of images captured for the whole project in each audit period**. That's **903 GB of images** for the seven years.

Consider one person is allocated to audit just the Anaheim fueling stations (212) only.

2 (MB) x 4 (count) x 212 (stations) = **1.7 GB of images for Anaheim each audit period** (and the person will need last year's **1.7 GB of images** either accessible on the web or offline).

Project 1

Scenario

Assumptions

How much/many?

Project 2

Scenario

Assumptions

How much/many?

How large are individual photo files?

Digital cameras, smartphones, and computers capture photos in different default formats and sizes. Check with your device manufacturer for the options that you can configure to increase photo quality and decrease file size. As a guide, the following table shows typical file sizes for one of the most common types of image files, JPEG, at common sizes.

JPEGs use lossy compression, which means colors that are difficult to distinguish with the human eye may be removed, and hard lines and edges may appear fuzzy. If the highest level of detail is required in your project photos, you may need to consider other file formats.

JPEG file sizes

Resolution (MP)	Dimensions (pixels)	Good for print size at 300 ppi		Approx JPEG 100% 24-bit/pixel file size (MB)
		Inches	ISO	
2	1800 × 1200	6 × 4	A6	0.4
3	2100 × 1500	7 × 5		0.6
7	3000 × 2400	10 × 8	A4	1.5
10	3600 × 3000	10 × 12		2.2
14	4200 × 3300	14 × 11	A3	2.8

Note: JPEG file sizes in this table were estimated using https://toolstud.io/photo/filesize.php.

How many map tiles do I need?

Imagery files have been used in the field for many years, and these files can be of variable sizes and resolutions. More recently, apps have standardized using map tiling schemes that break up the world into consistent sets of tiles at nominated zoom levels. This makes it possible to estimate how many tiles, and therefore how much memory, are needed to host that data. Use the following "Zoom Levels and Tile Memory" table to determine how many map tiles you'll need.

The exact amount of map tiles needed for a given area on the globe is dependent on the precise location and the amount of detail that is required to be viewed on the screen. The table includes columns from OpenStreetMap to describe what kind of geographic information can be seen at different zoom levels.

Zoom levels and tile memory

Level	Number of tiles	Tile width (° of longitudes)	~Scale (on screen)
0	1	360	1:500 million
1	4	180	1:250 million
2	16	90	1:150 million
3	64	45	1:70 million
4	256	22.5	1:35 million
5	1,024	11.25	1:15 million
6	4,096	5.625	1:10 million
7	16,384	2.813	1:4 million
8	65,536	1.406	1:2 million
9	262,144	0.703	1:1 million
10	1,048,576	0.352	1:500 thousand
11	4,194,304	0.176	1:250 thousand
12	16,777,216	0.088	1:150 thousand
13	67,108,864	0.044	1:70 thousand
14	268,435,456	0.022	1:35 thousand
15	1,073,741,824	0.011	1:15 thousand
16	4,294,967,296	0.005	1:8 thousand
17	17,179,869,184	0.003	1:4 thousand
18	68,719,476,736	0.001	1:2 thousand
19	274,877,906,944	0.0005	1:1 thousand
20	1,099,511,627,776	0.00025	1:5 hundred

Source: Zoom levels. (2023, January 18). OpenStreetMap Wiki. Retrieved 06:31, June 20, 2023 from https://wiki.openstreetmap.org/wiki/Zoom_levels.

Note: The table includes two columns on the right to help you estimate the amount of memory needed for two representative-sized projects: 2,500 sq km (about one square degree) and 100 sq km (about the size of Disney World).

<table>
<tr><th rowspan="2">Examples of areas to represent</th><th colspan="2">Approximate amount of memory required for map tiles covering an area of...</th></tr>
<tr><th>2,500 sq km</th><th>100 sq km</th></tr>
<tr><td>Whole world</td><td rowspan="10"><1 MB</td><td rowspan="13"><1 MB</td></tr>
<tr><td></td></tr>
<tr><td>Subcontinental area</td></tr>
<tr><td>Largest country</td></tr>
<tr><td></td></tr>
<tr><td>Large African country</td></tr>
<tr><td>Large European country</td></tr>
<tr><td>Small country, US state</td></tr>
<tr><td></td></tr>
<tr><td>Wide area, large metropolitan area</td></tr>
<tr><td>Metropolitan area</td><td>1 MB</td></tr>
<tr><td>City</td><td>4 MB</td></tr>
<tr><td>Town, or city district</td><td>14 MB</td></tr>
<tr><td>Village, or suburb</td><td>53 MB</td><td>1 MB</td></tr>
<tr><td></td><td>207 MB</td><td>3 MB</td></tr>
<tr><td>Small road</td><td>828 MB</td><td>8 MB</td></tr>
<tr><td>Street</td><td>4 GB</td><td>40 MB</td></tr>
<tr><td>Block, park, addresses</td><td>11 GB</td><td>115 MB</td></tr>
<tr><td>Some buildings, trees</td><td>100 GB</td><td>1 GB</td></tr>
<tr><td>Local highway and crossing details</td><td>400 GB</td><td>4 GB</td></tr>
<tr><td>A midsized building</td><td>1600 GB</td><td>16 GB</td></tr>
</table>

Cameras

Next to the paper notebook, cameras have been the predominant technology used in fieldwork across all industries for decades. Photos of wildlife, natural landscapes, building construction, and damage from flood, wind, and fire all immediately convey the subject matter of the study location.

Now that digital cameras are typically integrated into the devices we choose to use for our field GIS, the attachment of images to records of information is near effortless. But absence of integration hasn't stopped the collection of photos in the past, nor does it prevent someone using a dedicated camera for the task of documenting findings today. Splitting up technology can mean that you get to choose the best device for each task without compromise. Or the camera may simply be better off in someone else's hands because they are the better photographer.

When a courier delivers a parcel to your home and takes a snapshot of the parcel going into your mailbox to prove it was delivered, the integrated camera on the data capture device is more than sufficient to do the job. The resultant photo will be viewed in the pop-up of a web map or may be emailed as evidence.

When you are conducting scientific research of butterflies in the wild, a basic smartphone camera may not be your tool of choice: a tripod-mounted professional camera or fixed camera that is motion triggered may be more appropriate.

Descriptive photos no longer need to be taken from a device in the physical hands of a fieldworker—front-facing cameras can home in on distinct physical features, capturing detail that may previously have required a fieldworker to climb or scramble to a dangerous location to capture it.

Image quality from smartphone cameras has also greatly increased in recent years. So much so that, as a fieldworker, you will most likely want to downgrade the default photo quality options on your device. Downgrade? I can hear your shock. If you are taking hundreds of photos of trees, railway tracks, or fire hydrants and uploading them over a cellular network, considerations of (upload) time and (disk) space may far outweigh the number of pixels captured.

Knowing that typical computer desktop resolutions are only 1,920 × 1,080 or 2,048 × 1,152, the 7-megapixel images taken by many cameras exceed the resolution needed for displaying a good-quality picture on a computer. When planning to capture photos during fieldwork, you must consider where they will be used (on a desktop computer, printed on paper, or displayed in an app on someone else's mobile device) and how they will be transferred and stored.

Once you have chosen the best camera for the task and configured the default settings to always capture the best-sized images for your needs, what about the act of taking the actual photo? Do you need multiple photos of the subject from different angles? How much background (if any) do you need to include for context?

When the subject is a single person, animal, plant, or structure, following basic principles of portrait photography is useful: center the subject in the image, focus on the subject, stand close enough to capture the required detail, and move (or adjust settings) to get the best lighting.

If you are taking photos of a location, perhaps multiple subjects in their surrounding environment, it is wise to borrow just a little advice from landscape photographers. Frame the subjects to tell the story. If you are capturing a photo of a large tree that is precariously hanging over a road as part of a safety assessment, include the full extent of the tree and the full width of the road to represent the extent of the potential situation. Give a sense of scale. If you are recording findings in a tide pool on a rocky shoreline, as well as close-up subject photos, capture a wide view of the rock platform (preferably with a visible marker indicating the pool of interest) and include sea and land boundaries to show where in the wider landscape the pool of interest is. Is it close to the sea, where it will get a frequent exchange of water, or farther away where it may only receive a flush of water at particularly high tides?

Location photos taken for the purpose of fieldwork may not be destined to hang on a wall as art, so the rule of thirds, diminishing focal points, and leading lines may not be necessary, but they do need to tell a story that's clear to someone who wasn't there.

Which photo, which feature?

NAME: Marika | **ROLE:** Field technician
INDUSTRY: Water resources | **CIRCA:** 1999

In the late 1990s, before digital photography took hold, there was an evolution in photography that was just made for field data collection. Advanced photo system (APS) film was 24 mm, came in a cartridge that self-loaded, and allowed for the recording of user-entered metadata at the time of photo capture: photo title and number, exposure settings, film speed, and film ID. These subtle advances in technology made matching up information from paper or digital forms with the processed photos so much easier. Easier, but still not great.

I used these cameras while collecting data for the national river and coastal flood defense survey in the United Kingdom. Throughout winter (when the typically deciduous vegetation of the area was at its sparsest, and access to the riverbank was clear), I plodded along both banks of the River Mole and the River Medway, recording on my clipboard the presence and condition of walls, weirs, bridges, buildings, and vegetation that might contribute to (or abate) flooding in times of high rainfall. Sometimes I was on my own, sometimes I went with another technician. We would take photos with the (at the time) fancy APS camera and record the photo number on a paper form along with a sketch, measurements of size, and an assessment of condition. When working as a pair, one would use the camera to take photos and the other the clipboard to record the information.

However, when we returned to the office, the work was only just starting. Someone else was responsible for adding our written records and photos to the GIS. Risk of data transcription errors aside (my handwriting wasn't that bad), it was a lengthy process, but at least the numbered digital photo files that came along with the physical prints made the integration of photos into the database relatively seamless.

Fast-forward 20-plus years, and pictures taken with the cameras built into today's mobile computers and smartphones far exceed the quality of photos I took with that APS camera. Doing the same flood defense survey now would be far more streamlined, capturing the data and photos together on one mobile device. But I can still see the value in splitting the work among field team members, one with the camera and one filling in a digital form. The camera could be on another smartphone with the same app that can access the same record the other team member is using, or it could be a stand-alone digital camera. In the case of the stand-alone camera, a consistent method of photo enumeration, like the APS camera's, would be required.

Tools will change but good process stands.

From tape and compass to laser rangefinders

Not everything old is redundant. That may be a scary statement to minimalists or futurists, but bear with me as I explain.

Technology has come a long way, bringing people together from all over the world. Families spread across continents can video call and be together virtually. In well-connected locations, fieldworkers can transmit data collected in a city park directly to the head office, and a maintenance crew can immediately be sent a work order to go to the park and repair damaged playground equipment.

But consider being in dense, steep jungle 100 km from the nearest town, traversing a riverbed that is snaking its way up a small mountain. Your team is collecting samples and geophysical measurements and needs to geolocate each find. GNSS accuracy within the ravine is poor at best, and a satellite-based correction service (which could be cost prohibitive) may not even help with the poor reception.

In this situation, even in the early twenty-first century, ground-based measurement may be your best option. Explorers and surveyors have been using these techniques for thousands of years, and in some circumstances—like our ravine—they may still be the best option.

A tape and compass system is mostly self-explanatory. Starting at a known location—in this case, perhaps near where the river emerges from the dense jungle into an open plain where GNSS locations can be accurately obtained—you can use a compass to measure the direction of the next measurement and use a tape to determine the distance from that known location. Moving to that location and measuring the angle back to the first location (backsight) adds information to the calculations, and traversing from point to point in this manner can result in a network of calculated locations based on their relative position from the first known point. Of course, in the ravine, measurements also need to account for the changes in elevation, but this low-cost, low-technology option may be something to consider in these challenging environments.

Back in the connected world, there is plentiful scope for a similar measurement technology. A laser rangefinder can be used to capture the location and dimensions of features at a distance. This method of measurement is primarily for data capture in locations that either cannot be accessed or are not safe to work in.

Consider the capture of a row of trees along a vegetated median strip of a major road. Ideally, the location of these trees would have been recorded at the time of planting or captured from imagery. If these sources of information are not available, a field survey is a good way to establish the tree database. A median strip maintenance program is important to ensure that the vegetation doesn't impact traffic, but the act of data capture also shouldn't impact traffic or put the tree inspector in danger. By walking along the pavement on one side of the road and using a laser rangefinder, a fieldworker can accurately locate each tree and even measure its height without interrupting traffic.

Consider another example: in an underground mine tunnel, while standing at a known location at a safe distance from heavy machinery, the fieldworker can record the location and dimensions of a coal seam recently revealed on a newly excavated wall using a laser rangefinder.

The modern laser rangefinder uses the same mathematical principles as the tape and compass; only the precision of measurement, and perhaps ease of use, has improved.

Look out for the highway collectors

NAME: Alix | ROLE: GIS collector
INDUSTRY: Transportation | CIRCA: 2016

Field conditions can sometimes present precarious situations. Capturing data may not always be possible or safe when assets are in deep brush, next to hazardous materials, or along bustling highways. Tools such as laser rangefinders allow crews to collect data from afar, ensuring their protection while generating accurate locations.

While working for the Nova Scotia Department of Transportation and Infrastructure Renewal, I traversed the province's busiest highways. My crew was tasked with collecting street sign, streetlight, and culvert data to create a geographic database of highway assets. As you can imagine, walking along a highway is no joke, especially when some roads in Nova Scotia can see up to 25,000 cars each day.

My team wore bright protective gear and started work before sunrise so we would avoid heavy traffic, but cars can be unpredictable any time of day. One time when we were collecting data (without a rangefinder), a car nearly ran into me and my teammate. He had to grab my arm and pull me into a ditch. We were fine, albeit a bit shaken up. It just goes to show how every precaution must be considered when preparing for work in the field. Fieldwork isn't just tapping buttons and filling out forms; it's being aware of your surroundings and ensuring you're using all the safety equipment at your disposal. Given the haphazard nature of traffic and how it interfered with our work, safety was everything.

I spent a lot of my time hiking down the sides of highways collecting culvert locations, but street signs were a bit trickier to capture. They're located closer to the road (and thus closer to traffic) and are often found in conditions that aren't safe to walk through. That's why a buddy of mine used a rangefinder to capture this data. He'd walk along a safe stretch of land next to the highway and use the rangefinder to capture data from afar. All he had to do was point the rangefinder at the sign and collect the point. He did this up and down the Nova Scotia highways, creating a highly accurate database of the province's street signs.

This job enabled me to see so much of Nova Scotia—we collected so much data. So, the next time you're driving to work, think of all the points that were collected with each sign and culvert you pass—and be sure to watch out for the folks collecting them.

Rangefinder and other accessories log

Keep a log of the rangefinders and other accessories you have used, tested, or evaluated and save for future reference.

Make	**Model**	**Compatible OS**			**Proprietary app required? Name**
		iOS	**Android**	**Windows**	

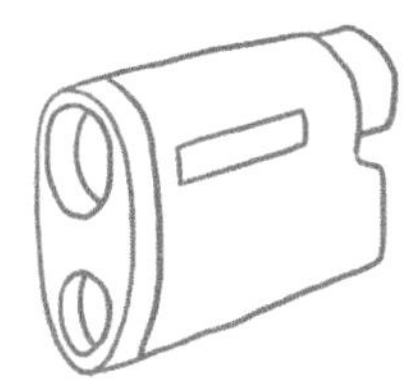

Comments

Accuracy, battery life, project suitability

Notes

4 Data

Location measurement

The geographic location of a feature in a GIS is its most recognizable element.

There's a little place in the Atlantic Ocean affectionately called Null Island by GIS professionals. It sounds mysterious, but this is the place where GIS features with no geographic coordinates are displayed. No geographic coordinates mathematically translates to coordinates of 0 degrees latitude and 0 degrees longitude, and that point is located off the west coast of Africa. There is no physical land mass at this place—it's an imaginary island—but when you open a GIS and find a thousand points located at 0, 0, something is clearly wrong with the location measurements of your data.

When sitting in front of a desktop GIS, you can move points and vertices around the map, relative to other points, vertices, and pixels, to estimate the location of a feature. But a GIS is a representation of the outside world, so a true measurement of a feature's location in the real world is a fundamental action performed in fieldwork.

People have been measuring places on Earth for thousands of years. Measurements from recognizable physical landmarks can produce a local reference system, and sometimes this is all that is needed. The precise location in the whole wide world may not be required for a fuel storage tank, but the fact that it is 10 m from adjacent buildings and dangerous equipment may be critical.

Every day many of us use our phones to show our location or share it with others. We may use an app to navigate to the one hardware store that stocks the exact-size bucket we need, or we may submit a geolocated photo of rubbish on the sidewalk to the local government authority to request a pickup.

Phones use a combination of hardware and software to determine your location—and depending on that combination, the accuracy and precision of that measured location can vary significantly. For our everyday use, an accuracy of 20 or 50 m is more than enough. By the time you approach the driveway of the hardware store, you have already forgotten about your navigation needs, and the pile of rubbish only needs to be located to the nearest neighboring address for the garbage truck driver to find it.

However, when you are deciding where to dig a trench for a new pipeline or ensuring that you are cutting branches from the correct tree, it's important that you're on the right side of a property boundary. Accuracies of 2 or 5 m are most common in field GIS, but sometimes submeter accuracies are essential.

Now is a good time to take a step back and revisit some terminology, starting with those terms: *accuracy* and *precision*.

Accuracy is used to describe how close a value is to its true value. The location report from your phone might look pretty good, but take a close look at that big circle that surrounds the point. That's the representation of accuracy, and it can easily represent a 100 m radius from the calculated location.

Precision is how replicable a measurement is. Every day, when you sit at the traffic lights at the same time, your location is going to show the same. Again, in the case of automobile navigations, 100 m is not likely to concern you. But if you are capturing the location of where a service manhole is to be dug, and you will be sending someone else to dig the hole, you will want that location measurement to be repeatable.

When choosing a tool for measuring location, you must be clear on what levels of accuracy and precision are needed for any given task. Either can be significantly affected by the environment and hardware and software choices.

Your phone has a GPS in it, right? It says so on the device specification. Phones usually have some variation of GNSS receiver hardware, but typically they rely on other sensors in the phone to provide you with fast and useful locations. This doesn't mean you will have the most accurate and precise

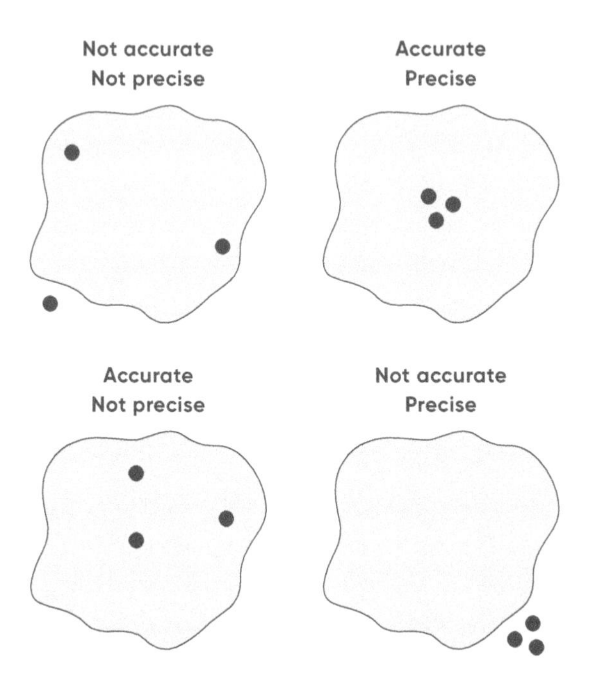

Four views of a tree with point representations of measured center location varying in accuracy and precision.

locations, but your phone battery will be preserved to a reasonable level of power consumption, and you will see a reasonable representation of your location at all times.

A common way that phones improve their location measurements is to use the network of cellular towers that they already use for communication. Each cell tower has a known location, and the signal strength (and unique identifier) received from at least three towers to the phone can be used to approximate the location of your phone. Because the device is already receiving this information to do its job as a phone, it is an efficient way to supplement the work of the GNSS hardware in the device to get a quicker location measurement.

When accuracy and precision are important to you, the steady stream of data messages from positioning satellites is best received and interpreted by a dedicated GNSS receiver. These National Marine Electronics Association (NMEA) messages contain everything needed to calculate a location with measurable accuracy.

Your chosen field app should offer options to get the most from your GNSS receiver. There should be options to prohibit data capture (or at least warn you) when preferred location accuracies are not met. There should also be the ability to capture location metadata so that you can later confirm the accuracies and precision of location capture.

Every last fossil

NAME: Jane — ROLE: Paleontologist

INDUSTRY: Paleontology and geology — CIRCA: 2008

More than 34 million years ago, there was a popular watering hole in an area now known as Badlands National Park in South Dakota. The watering hole nourished many prehistoric animals, including a giant piglike creature known as *Archaeotherium*. Fast-forward to 1997, when the first fossils were discovered at this site, launching the decade-long excavation known as the Big Pig Dig. I worked on this project during one of the 15 field seasons it took to excavate every last fossil from the site.

My job was to excavate and identify fossils, and then help measure the exact location and perimeter of each fossil before it was removed from the site. These measurements would then be used to create a 3D map. This meant measuring multiple x-, y-, and z- (elevation) coordinates for each discovery. The resulting map showed the location and shapes of all the fossils and charted the accumulation and change of fossil material in the watering hole over time.

To measure the location of each fossil, we used a surveying instrument called a total station. A total station uses a series of laser measurements to collect precise geographic information. For us, that meant elevation and the exact perimeter of each fossil. It was a meticulous process—as we made our way through the pit excavating dirt and rock, we would measure every fossil we came across. The total station would be calibrated once per day, and the perimeter of the pit captured. As each new fossil surfaced, its own measurements were captured with the total station. Multiple people would work in the same pit every day, side by side, uncovering fossils.

The year I worked at the Big Pig Dig, we were working at a depth of about 4–5 ft, and we excavated 299 pieces of fossil material. In the 15 field seasons it took to complete this project, more than 19,000 fossils were recovered and mapped. This feat would not have been possible without the total station and the paleontologists who used it to measure the precise location and perimeter of each fossil—creating an incredible 3D map and preserving paleontological history in the process.

Location measurement terminology

The following table lists key location measurement terms.

Term	Description
Accuracy	How close a value is to its true value. Can be influenced by environmental conditions (tall building and tree cover creating interference) or the quality of the GNSS receiver and software used with it to calculate the resultant values.
Precision	How replicable a measurement is. Mostly influenced by the hardware.
GNSS	Global Navigation Satellite System. General term used for system of satellites used to calculate position measurements. Can incorporate satellites from different constellations.
GPS	Global Positioning System. USA constellation of satellites. The first and mostly widely used system, the acronym is commonly used to describe GNSS technologies in general.
AGPS	Assisted GPS. Term used in consumer-grade device specifications to denote an assisted location measurement system. Typically, these systems use cell phone tower triangulation and software estimations to get faster position measurements and minimize battery consumption.
DGPS or DGNSS	Differential GPS or GNSS. Uses land-based reference stations to broadcast differences between positions determined by satellite systems and known fixed positions.
Differential age	Age of the differential signal and correction used by the GPS receiver to differentially correct the position.
GLONASS	Russian constellation of satellites.
BeiDou	Chinese constellation of satellites.
Galileo	European constellation of satellites.
IRNSS	Indian constellation of satellites.
QZSS	Japanese constellation of satellites.

PPP	Precise Point Positioning. Calculates precise positions. Similar to DGNSS but uses only a single receiver.
RTK	Real-time kinematic positioning. Used to enhance the precision of position measurements. Requires one or more nearby accurately surveyed reference stations.
SBAS	Satellite Based Augmentation System that broadcasts DGPS correction messages from geostationary satellites to improve position measurements from GNSS.
WAAS	Wide Area Augmentation System. SBAS system servicing North America.
NMEA	National Marine Electronics Association. An NMEA sentence received from a GNSS contains all the information about a given position.
HDOP	Horizontal dilution of precision. Measure of possible error in horizontal positional measurement precision.
VDOP	Vertical dilution of precision. Measure of possible error in vertical positional measurement precision.
PDOP	Position dilution of precision. The equation used to determine PDOP is $PDOP^2 = HDOP^2 + VDOP^2$.

Record additional terms that you come across that are important in your industry or projects.

Term	**Description**

Notes

Tracking

In the popular press, tracking workers has a bad name: Big Brother is watching you, employers do not trust employees, and privacy is nonexistent. But near real-time location tracking can be useful for identifying where field crews are at a given time, to report project progress or to optimally move fieldworkers to different tasks or sites. Also, when fieldworkers are working in remote locations, their personal risk and safety management needs may quickly outweigh any hesitancy they might have about privacy.

As with almost everything, a fine balance must be sought. In some environments, enabling tracking only while working within a specific app or project means that when fieldworkers stop for lunch, their precise location will not be recorded.

In GIS field apps, tracking could be achieved in several ways:

- A fieldworker could explicitly capture a line, or collection of continuous points, that represents where they traveled.
- An app could automatically capture continuous points—and generate a single line from them—that represents where the fieldworker traveled.
- An app could continually update the last known location of the fieldworker, a single-point record.

A fieldworker explicitly capturing a line that represents where they went is good as a historical reference but not useful for real-time tracking. It allows the greatest amount of freedom—the fieldworker can start and stop the line when they want—but for this reason, it can be the most unreliable and accident prone, and easily forgotten. This line is no different from any other line in the field GIS; it can have attributes that the fieldworker can fill in to annotate the line.

When an app can continuously capture points and present it as a tracklog, it typically doesn't prompt the fieldworker with a data entry form for completing attribute information (as in the case of explicit line capture), and it can carry on in the background while the fieldworker is doing their work. This unobtrusive tracklog capture is useful when coordinating teams of fieldworkers needed to cover a designated area of ground. For instance, a live view of the tracklogs of search and rescue teams overlaid with a point layer representing findings can have a command center immediately redirect teams to untracked locations or locations nearest to the finds.

In the case of safety and risk management, it may be enough to have just the last known location reported by a fieldworker's device. Synchronizing the geographic location of this single point record at a regular time interval to a central database can enable the creation and display of a dashboard showing the near real-time movements of fieldworkers. If these last known locations are not updated at the expected interval, alerts can be raised and fieldworkers contacted to establish their well-being.

Relevance of tracking

Consider the following questions when deciding whether tracking is needed in your field GIS and whether real-time or historical tracking records are useful to you:

- Is your fieldwork being conducted in a hard-to-reach location?
- Is your fieldwork being conducted in a dangerous location?
- Do you have multiple workers or teams of workers in the field that you could redeploy based on real-time information?
- Do you need to maintain records for regulatory purposes of where and when fieldwork was conducted?

- Do you need a tracklog of all locations visited (a collection of points, breadcrumbs, or a continuous line), or is the last known location enough?

The "Type of Tracking by Industry" table lists some industry examples of tracking. In this table, the following terms are used to differentiate types of tracking:

- **Historical:** Tracklog that is captured on the device and shared or reviewed later. It is a historical record of tracks.
- **Real time:** Tracklog (or last known location) that is sent in real time to a database or dashboard for immediate view by others.
- **Foreground:** The fieldworker is actively starting and ending the capture of their tracklog, and they can annotate the record with additional information such as a name or description.
- **Background:** The fieldworker is not actively capturing the tracklog. It is done in the background, and no additional information can be annotated to the tracklog.

Type of tracking by industry

Type of tracking	Industry examples
Historical foreground line or continuous point capture	**Agriculture:** Capture a transect (line) along which observations are recorded (points) that document the identified species at that location. **Natural resources:** Capture a line that represents a length of riverbank (either walked along the bank edge or followed by boat on the river) that was surveyed for potential erosion hazards. **Public health:** Capture a line that represents mosquito treatments applied by truck spraying.

Historical background tracklog	**Population census enumeration:** Validate that census record capture was conducted on-site at the nominated address. **Asset inspection:** Validate that time spent at an inspection location satisfies minimum requirements for completing a suitable inspection. **Energy utility:** Maintain documented evidence of completed maintenance activities.
Real-time foreground line or continuous point capture	**Mining:** Capture boundary (with height) points from a drone that can be used to calculate size and slope of spoil heaps for safety management. **Fire service:** Capture a live fire boundary from a helicopter for use by ground crews for situational awareness. **Railroad construction:** Capture a line that represents real-time construction progress.
Real-time background tracklog	**Search and rescue:** Analyze in real time the movements of a search event and use this information to direct teams to the best location to search immediately. **Conservation:** Ensure optimum coverage of an area during an invasive species removal activity. **Science:** Monitor the location of sole fieldworkers in dangerous environments from a base camp.
Real-time last known location	**Transportation:** Deploy nearby road safety teams to an accident location immediately to manage traffic conditions and facilitate swift access to the site by emergency service vehicles. **Emergency response:** Deploy relevant and nearby personnel to an incident from a command center. **Major events:** Redeploy nearby event personnel to a location from an event coordination center.

Coordinate systems

The topic of coordinate systems can scare many, but some knowledge of the subject can save a lot of grief in fieldwork. At the simplest level, a coordinate system is an organized way to represent points in space.

A geographic coordinate system is used to define locations on a spherical representation of the earth. Latitude is the measure of the north–south position on that sphere, and longitude is the measure of the east–west position on that sphere. And when I use the word "sphere," it's more like a squashed orange than a tennis ball.

A projected coordinate system is used to represent locations on a planar representation of the earth. Enter the paper (or digital) map.

What coordinate system does your field GIS use? This may be dictated by your device and app of choice, or it may be something you can customize. What coordinate system does your GNSS hardware use? It may be a default or something that can be customized and may be different from what your field GIS will use. Now you can see why a little knowledge about coordinate systems is useful.

Internet mapping typically uses the de facto standard geographic coordinate system called Web Mercator—also known as Google Web Mercator, Spherical Mercator, WGS84 Web Mercator, and Pseudo-Mercator. Coordinates are represented as latitude and longitude, and apps that display web maps will usually use this coordinate system by default. This coordinate system makes a good representation of area and distance around the equator but produces distortion as you move away from the equator.

Fieldwork tends to be localized—for someone working solely in London, or anywhere else far from the equator, there may be little (or no) benefit in using a coordinate system such as Web Mercator. For our London worker, the British National Grid may be more suitable. This grid is discrete and detailed in its coverage. Covering the landmasses of the islands of Great Britain, each National Grid reference is composed of two letters and up to 10 digits to represent a location to the nearest square meter. Coordinates are measured in eastings and northings and have units of meters.

By default, GNSS uses the WGS84 geographic coordinate system, delivering values of latitude and longitude for measured locations. Can our London-based fieldworker use GNSS? Yes, but their receiver, or app, may need to transform the coordinates from one system to another. Some receivers may perform the additional calculations and adjustments needed to match a local coordinate system directly. This is worth researching before choosing a receiver off the shelf.

If you are working in relatively flat locations, calculations and adjustments in the north–south and east–west directions alone may be enough. But consider capturing data in the field to build a topographic map for the country of Switzerland. This mountainous country is poorly represented by the "squashed orange" generalization of the WGS84 coordinate system, so a more complex local coordinate system should be used.

You need to know what coordinates you're capturing, what coordinates your GIS is using, and whether you can transform between them. Typically, these transformations can happen in the office on desktop computers when synchronizing data to source databases, but if real-time transformation is needed in the field, be sure to use equipment that can do this on the fly.

Coordinate system terminology

The following table lists key coordinate system terms.

Term	Description
Longitude	Longitude values are measured relative to the prime meridian (Greenwich, UK). They range from −180° when traveling west to 180° when traveling east, measured in degrees.
Latitude	Latitude values are measured relative to the equator and range from −90° at the South Pole to 90° at the North Pole, measured in degrees.
Geographic coordinate system (GCS)	A reference framework that defines the locations of features on a model of the earth. It is shaped like a globe—i.e., spherical. Its units are angular, usually degrees.
Spheroid	A sphere flattened at the poles that represents the earth's surface.
Projected coordinate system	Converts a GCS into a flat surface, using math (the projection algorithm) and other parameters. Its units are linear, most commonly in meters.
Datum	Part of the GCS that determines which model (spheroid) is used to represent the earth's surface and where it is positioned relative to the surface.
Geocentric datum	An earth-centered, or geocentric, datum uses the earth's center of mass as its origin. The most widely used datum is WGS84.
Local datum	A local datum aligns its spheroid to closely fit the earth's surface in a particular area.
Projection	The mathematical algorithm that defines how to present the round earth on a flat map. Common examples include Mercator, Lambert conformal conic, and Robinson.
WKT	Well-known text is a string that defines all necessary parameters of a coordinate system.
WKID	A unique number assigned to a coordinate system.

Geoid separation	Difference between the WGS84 earth ellipsoid and mean sea level as reported by the GNSS receiver. This is sometimes referred to as orthometric height.
Transformation	The calculations that convert your geographic coordinates (latitude and longitude) from one GCS to another so they will draw in the correct place. Sometime referred to as geographic transformation or datum transformation.

Record additional terms that you come across that are important in your industry or projects.

Term	**Description**

Notes

Basemaps and operational layers

A GIS consists of layers of geographic data. Information can be analyzed and queried across layers, creating new information and providing countless opportunities for new discovery and information creation.

Supporting a distributed, connected, and always current GIS for everyone to use requires significant effort to maintain the relationship between data elements and carefully manage versions of information.

Consider two field crews working to maintain a power pole network, replacing aging or damaged poles. One crew is instructed to work on the southeastern side of town, the other on the northeastern side of town. One team has had less damage to repair, so they are powering through the task. In the location where the two service areas overlap, the pole conditions are good, so there are no visible changes on-site for either team to know that the other has been there already. They both end up inspecting the same 10 poles on one street. How the GIS is designed can prevent this from happening in the first place, or at least provide a mechanism for the duplicate records to be resolved.

To reduce the complexity of a field GIS—and prevent costly mistakes and duplication of effort—separating layers into the two broad categories of basemaps and operational layers can be useful.

A basemap contains layers of information that are for reference only. The basemap may comprise raster tiles that show many layers of data flattened to a single read-only layer. Raster basemaps are typically fast to draw on the screen and pack a lot of visual information in a small file. A vector basemap also shows many layers and can be even faster to draw on the screen, but unlike the raster basemap, it can have scale-dependent rendering—so when you zoom in close, labels can reposition themselves for better reading.

Operational layers are layers that you may be editing. In our power pole scenario, these would be the layers that represent the poles, conduits, wires, and associated equipment being managed.

Another good way to think about your layers is to ask how often the information will be changing. Basemaps typically don't change for the course of a project. For instance, at the beginning of the mosquito abatement season, you may configure a basemap that contains road and property boundaries, polygons that represent last season's spraying locations, and other critical landmarks. Saving all these layers together as a basemap means you can deploy it once to all your teams' field devices and then leave them alone for the rest of the spraying season. Each field device may be synchronizing only one or two layers representing this season's spraying locations. This would mean that minimum computing power is required to send and receive updates to that layer. The teams can see the work of other teams (once they synchronize) and can't accidentally edit previous years' records.

The operational layers can be further managed, allowing specific users to perform different actions. The members of the spraying team may be able to add only new records, but the fieldwork supervisor may be able to edit existing records to perform quality assurance checks. All users would still be using the same basemaps, but their task-dependent data layers would differ.

By the stream, next to the tree

NAME: Gordon **ROLE:** GIS intern
INDUSTRY: Parks and recreation **CIRCA:** 2019

Before the dawn of modern GIS, the King County Parks system in the state of Washington used as-built drawings as a record-keeping system for thousands of assets across hundreds of parks. As with any paper inventory, it became difficult to maintain these assets and understand where they were.

Several years ago, the park system began creating a geographic database for all its assets so that no park bench or trash can was left unaccounted for. This is the project I contributed to during my two years interning with King County Parks.

Every day during the summer, I would go out with two or three other interns to one of the parks. Using our mobile devices, we would capture the location of various assets and take photos if we noticed any damage (damaged assets would feed into our work order system for our maintenance crew to come and fix them). We would also do some ground-truthing using the as-built drawings to verify that assets were where they should be. Sometimes they were...sometimes they weren't. Occasionally, we would also use high-accuracy GPS receivers to collect more exact locations. These receivers were attached to our hats, which garnered more than a few glances from Washington park-goers. Do they not know that the higher the receiver is, the better the signal?

We didn't always have access to these receivers, and the GPS on our phones wasn't always reliable. In these instances, we would rely on the basemap to cross-reference where an asset was in front of us to where its point should be placed on a map. If our GPS location (the blue dot) wasn't lining up with where the asset was, we could identify a stream or tree using the satellite imagery of the basemap and use that as reference for adding the asset to our database. So, when other park employees need to find these assets on a map, they'll be able to use the same reference points on the basemap to locate them.

Network datasets

In some industries and projects, it's critical that the points and lines in the GIS have an inherent knowledge of each other. By creating a relationship between adjacent features, network datasets can be used to model street traffic, the movement of gas or liquids in pipelines, and the environmental flow of water in stream and river systems.

As an example, think of the street closest to where you live. Imagine walking along the street until you reach an intersection. Which way can you go? Left? Right? Straight ahead? Perhaps you cannot turn left because the street that runs perpendicular to yours is a one-way street. Identifying the one-way direction of that street in a GIS means that the network dataset can be used to model how travelers can move around the transportation network.

When you use a network dataset, you can use your GIS to answer the following types of questions:

- What is the quickest way to get from point A to point B?
- Which incidents can be reached within five minutes of a team of responders?
- Which is the nearest inspection location from my current location?
- Which valve (or valves) do I need to close to stop the flow of water in the pipeline at a given location?

If your fieldworkers are using a field GIS to guide the shutdown of a portion of a sewerage network to enable repairs, an app that supports a network dataset can be a useful tool, for both the team in the field and those back in the office managing the full network and keeping customers informed.

Not all field GIS apps will support or maintain the network connection information of a dataset, nor do they need to. If your fieldworkers are performing a visual safety audit of stationary assets, the relationship between adjacent features may not be needed for that task. A maintenance crew working along the length of a cross-state gas pipeline clearing fire hazards—vegetation, dumped rubbish, damaged fences—may find it useful to see points along the line that represent valves or junctions for the purpose of locating themselves relative to the infrastructure (for example, "cleared a fallen tree 10 m southbound from valve number 239832"), but knowing the pipeline segments on either side of that valve is of no use. If the app, map, or form of choice doesn't break the relationships in the underlying GIS, different teams should be able to use the same data in different ways, allowing greater choice to pick an app, map, or form that's fit for their purpose.

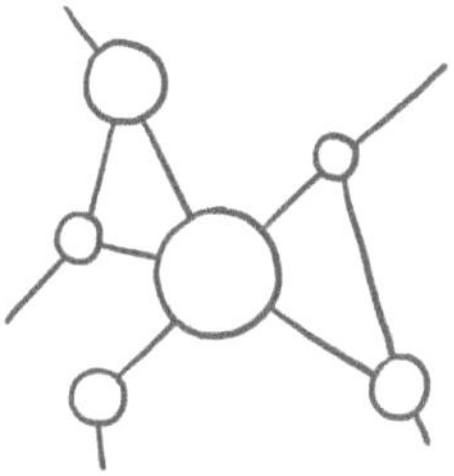

Uniquely identifying records

Sophisticated naming and numbering conventions have been used for decades to uniquely identify field observations, even those recorded on paper. An ordered combination of notebook, activity name, location, and date-time identifiers entered as a unique value can ensure that multiple team members can capture data on a common project. This also allows all researchers on the project to readily identify the context of all records.

Introducing a digital database, and more importantly, a distributed digital database into your fieldwork means that you need more careful record identification. A distributed database is designed for many users to synchronize edits at any time and include related images and documents. It will ensure correct metadata attribution for each addition, update, or deletion of information.

A GIS can use many ways to relate records from one layer to another. Sometimes this is done at the application level only: a spatial or attribute query is performed on the fly and reveals relationships that may not be defined in the underlying database. This provides great flexibility for someone interrogating data and can enable the discovery of new relationships that had not been intended in the original data capture. But when designing a database that will receive contributions from many people using many types of data, it's critical to define relationships in a way that can robustly receive and organize that data for later interrogation.

At first thought, it is tempting to just sequentially number records in the database: 1, 2, 3.... This is OK when you have a single table of information and one person capturing data. But what happens when you have two people capturing data, and they both have a record numbered "1"? Which record is which? A database will automatically include a sequential number for records in a table. In ArcGIS, a common GIS, this is the field called ObjectID, which is a unique number for the records in a feature layer or table. When new records are added, the next sequential number will be assigned to the record. If a record is coming from a replicated version of the database (for example, an entry coming from an app in the field that has a connection to the central database), the ObjectID will be populated in the parent database using the next sequential number, and the ObjectID for the same record in the replica (in the app) will be changed to match the newly synchronized value, as shown.

Before synchronization

Parent geodatabase

ObjectID	UniquePoleID
1	Pole1a
2	Pole1b
3	Pole1c
4	Pole1d

Child (field app) geodatabase

ObjectID	UniquePoleID
1	Pole2a
2	Pole2b
3	Pole2c
4	Pole2d

After synchronization

Parent geodatabase

ObjectID	UniquePoleID
1	Pole1a
2	Pole1b
3	Pole1c
4	Pole1d
5	Pole2a
6	Pole2b
7	Pole2c
8	Pole2d

Child (field app) geodatabase

ObjectID	UniquePoleID
5	Pole2a
6	Pole2b
7	Pole2c
8	Pole2d

The ObjectID can change after synchronization.

So, although conceptually the ObjectID is a unique identifier for the record in one table, when viewed across a database that allows synchronization of data, that identifier can and will change.

In practice, database design may include additional fields that maintain the relationship between records, including related tabular data, photos, and attached documents. The following figure shows a simplification of this synchronization.

If a human-readable unique identifier is useful in your database, concatenation of a series of short values can achieve the uniqueness required for a robust relational database. This can also be useful if you are using a combination of data collection tools—a mobile device, a paper notebook, and a scientific data logger that can't connect to your mobile device or computer in the field. For example, you could concatenate values for project area code, user ID, and date-time (to the second) for a human-readable code. Write the project area code and username on the front of your notebook and date your pages. Now your GIS has an accessible relationship with the contents of your paper notebook.

For example, the identifier Area123User456_20220621130530 is made up of the following:

Area ID	User ID	Date	Time
Area123	User456	20220621	130530

To guarantee a unique identifier for all records in your data, use of a globally unique identifier (GUID) is recommended. A GUID is a 128-bit integer (16 bytes) that can be used across all computers and networks wherever a unique identifier is required. Such an identifier has a very low (in practice, near zero) probability of being duplicated. Here's an example:

2bc4dae6-15d0-4533-ac50-0b99418a9ede

Before synchronization

Parent geodatabase

ObjectID	UniquePoleID
1	Pole1a
2	Pole1b
3	Pole1c
4	Pole1d

Related photo records

ParentID	Photo

Child (field app) geodatabase

ObjectID	UniquePoleID
1	15d0-4533...
2	92b3-4e7d...
3	c031-4b56...
4	668e-457b...

Related photo records

ParentID	Photo
15d0-4533...	File1
92b3-4e7d...	File2
c031-4b56...	File3
668e-457b...	File4

After synchronization

Parent geodatabase

ObjectID	UniquePoleID
1	Pole1a
2	Pole1b
3	Pole1c
4	Pole1d
5	15d0-4533...
6	92b3-4e7d...
7	c031-4b56...
8	668e-457b...

Related photo records

ParentID	Photo
15d0-4533...	File1
92b3-4e7d...	File2
c031-4b56...	File3
668e-457b...	File4

Photo records are related to objects through the objects' GUID in the synchronized geodatabase.

By assigning all data captured in the field (and all the data in your database) a GUID, record synchronization can methodically join records that are related to each other without adding any unintentionally duplicated identifiers. Returning to the replicated database scenario just described, imagine that each new record collected in the field also has several photos attached to it. On the mobile device, the new record has a GUID to identify itself, and each photo references that GUID. When the record is synchronized with the parent database, it may adopt the next sequential ObjectID in the database, but its GUID will remain—and in fact be inserted into the database. The photos will also get synchronized to the parent database and maintain their relationship with the field record, no matter where it is accessed.

The devil is in the details

NAME: Philip ROLE: GIS specialist
INDUSTRY: Environment and heritage CIRCA: 2016

Among the vast number of natural landscapes managed by the state government parks and recreation organization I work for, there are also more than 150 heritage sites. They're scattered across parks, sanctuaries, and reserves, and many are included on the state heritage register. These sites contain rich histories that are of great cultural significance to local communities, so it's especially important that they be preserved for future generations. The state government recognizes their importance by requiring maintenance and inspections of each site.

For more than 50 years, the historical preservation efforts of the organization were managed with paper forms—the meticulous inspections of every window, door, and fixture written down in a never-ending, unsustainable paper trail. This presented the perfect opportunity to digitally modernize their workflow with GIS.

As a GIS specialist, I worked with the heritage team to replace its paper forms with surveys that could be accessed on a mobile device. The adoption was quick. Team members took to the field with their devices, creating the first geographic database of heritage site inspections using digital forms.

When the time came for the second iteration of inspections, it became clear that a relational database structure would significantly improve results and in-field workflows. Rather than have multiple records for each site (one for a window, door, and so on), I helped the organization transition to a system of related records. This means there would be only one record (and point on the map) for each heritage site. Within that record, there would be a related record for each element of the site (window, door, and so on), with each element having its own inspection records to represent each annual inspection.

After integrating this relational database structure, team members could view the entire history of a site's inspections by locating and viewing the information related to one record in the field. A full to-do list of inspections could be provided, ensuring no elements were missed, and reports could easily be generated for specific groupings of records.

Just as the history of the heritage sites themselves is important, so too is the history of their condition and inspections. By moving from paper forms to a digital, relationship-based database, the organization now has infrastructure to help preserve these sites in an efficient and sustainable way.

Photos and other documents

Carefully managing and storing photos is an extremely valuable element of a GIS. But a quick look at the photo folder on your own smartphone will show you how file sizes and file numbers can quickly multiply.

Your own phone may do its own photo management. Some apps may automatically downsize a selected photo before sending it as part of a message, or your backup app may choose to wait to send your photo to the cloud until you have a Wi-Fi connection. These settings are all designed to minimize data traffic and improve the performance of your device.

The same considerations should be made when including photo capture and photo viewing in your field GIS.

If your fieldwork task is a condition inspection and you need to compare the current condition of a building façade to its condition from last year's inspection, it is important to have not only the maintenance records for the last year on hand but also all the photos from last year's inspection for comparison.

Imagine that you are the heritage officer for a small local government area, and you need to complete this heritage survey for the 200 buildings in your town's central business district. Let's assume that each building will have approximately five high-resolution photos, and each is approximately 3 MB. That's a total of 3 GB of images to download to your field device before you can start this year's assessment. Maybe you should download those images in the office before you head out into the heat of the day.

For other projects, it may be enough to capture a photo and submit it to the GIS. Previous photos do not need to be downloaded to the field device, and you can wait to send today's photos until you are back in range of a good internet connection.

However you choose to download and upload your photos, it's critical to ensure they have a connection to your GIS. A uniquely identifying GUID is the most robust method of identification but writing a description in a paper notebook with a date and time stamp can often be enough to appropriately attribute the photo later.

Photos are not the only type of file that can be attached to a field GIS record. Other files that you may consider attaching to a record include the following:

- Access permit PDF that was received by email while on-site
- Screenshot PNG file of the mobile device map with hand-drawn graphics on it
- Data file CSV that comes from a sensor and can be analyzed later

Each of these types of files can be saved locally to a field device and shared, moved, or managed later, but by adding them to your GIS you can significantly improve their contextual value. An artifact is nothing without context, and if you prefer to be the archaeologist (who treats an artifact as knowledge) rather than the treasure hunter (who treats an artifact as bounty), carefully organizing these kinds of files can uncover opportunities for better data management, new ways to share information, and new discoveries.

Machine learning

Is there a place for machine learning in fieldwork? Of course.

Will it completely replace fieldwork? Not likely.

The terms *machine learning* (computer models built from sample data), *deep learning* (multiple layered models), and *artificial intelligence* (machines performing tasks usually requiring human intelligence) are common in modern technologyspeak. These terms are all very cool and exciting for most people, up until the media starts reporting that department stores use facial recognition tools to track customers, and Big Brother is supposedly watching from everywhere. At that point, we get a little concerned.

During the Industrial Revolution in the late 1700s, hand production methods shifted to machines, and the mechanized factory system significantly changed the way people worked. People were not made redundant, but they were freed up from their manual labors to create new tools, products, and processes that made life more comfortable and productive.

Similarly, the digital revolution of the mid-1900s shifted mechanical and analog electronic technology to digital electronics. Again, significant improvements in efficiency and volume of outputs were achieved, but again, human workers were not eliminated completely—only their role in the workforce changed. Through the automation of repetitive tasks, safety improved in the manufacturing industries, as did quality and quantity of production.

If the next technological revolution is of neural networks (a computer system modeled on the human brain and nervous system), we must embrace the opportunities that come with it. By incorporating machine learning into fieldwork tasks, we can focus the activities of the fieldworker on the scientific research, asset maintenance, or situational awareness activities that they are there to accomplish.

Most commonly, fieldwork is, in fact, an input to desk-based GIS research projects, used to validate data that has been generated by models analyzing aerial photography or satellite imagery against ground truth. Validating the output of these kinds of models in the field enables the reliable extrapolation of data at a regional, national, or even global scale.

Through the careful choice of an app with machine learning capabilities, a data capture project becomes a data validation project that can be completed quickly.

Another set of eyes on the road

NAME: Someone ROLE: Maintenance technician
INDUSTRY: Roads and highways CIRCA: 2030

The following story is fictitious. It is a 2023 perspective on what fieldwork could look like in 2030. Each of the elements of technology described already exists but is used either on its own or as part of trial projects. By 2030, this kind of combination should be the norm.

Picture this: My team and I work hard to keep the road networks in our city safe and free flowing. Whether we are conducting repairs to pavement, clearing accident sites, or performing maintenance activities, our daily tasks are delivered to our truck's computer in real time and in the order and frequency that we can best complete them.

The computing system also keeps track of our onboard inventory of personnel, tools, and materials, so when we return to the depot, the items that we need to restock are ready to go, and tools that need maintenance or replacement have been identified.

Going to and from jobs obviously involves our own time as users of the road network. While we are driving, we make use of another set of eyes to collect real-time data for an ongoing maintenance project.

Street signage—stop, yield, merge—is typically installed to provide information to drivers, and the retroreflectivity of a sign is a measure of people's ability to see the sign, particularly at night. In our jurisdiction, minimum retroreflectivity standards must be met, so street sign inspections are an important maintenance activity. In the past, we would deploy teams of people to drive to every sign, pull up next to it, then fill in a digital form confirming the location and asset ID of the sign and recording its retroreflectivity. Using this process, it would take at least several minutes to capture each record. Across our city, that would accumulate to several weeks' work for each complete run.

Now we have a set of cameras and retroreflectivity instruments mounted on our truck, and as we drive between jobs, photos and measurements are automatically captured when we are near a known street sign. The computer can analyze the image and create a retroreflectivity measurement record for that sign at that time. Signs whose retroreflectivity falls outside the recommended range are automatically added to an inspection task listing. Also, compliant signs that we drive past often play a role in refining the model; photos taken in different lighting and weather conditions are added to the model to help improve it.

Of course, we don't pass every sign in our city in one day's or one week's or one month's work, but we do minimize the number of locations that a specialized project team needs to visit during targeted retroreflectivity assessment projects. And our repeated capture of sign images continually improves the quality of the tool used by these teams.

With this app, automobile drivers can photograph the sign as they drive past it. The location can be extracted from the metadata of the photo, and the sign type and retroreflectivity can be automatically deduced from the photo by processing it against an image classification model that was built from thousands of photos of differing type and retroreflectivity. The classification predictions are automatically added to the data entry form but can be refined or corrected by the fieldworker if needed.

Welcome to the future!

Results on the run

Whether you have 10,000 enumerators in the field collecting census data, a dozen crews maintaining a city's sewer and water network, or just two people inspecting grapevines for pests, it's critical—or at least highly useful—to be able to see the fieldwork unfolding in real time.

In the case of grapevine workers walking up and down corridors of vines, it may be enough that they can synchronize their tracklogs and inspection information continually as they work and see one another's progress on their own devices. This would help ensure they don't overlap their work and don't miss important sections.

Synchronizing the work of sewer and water field crews would not only allow other teams within the operation to immediately see the results on a dashboard to track completion, but the data changes could automatically trigger adjustments in the operation of the network. To replace a broken valve, a portion of the network may have been shut down. The replacement of that valve and information updated directly into the GIS could automatically initiate a system operation that would reactivate that portion of the network.

To complete a population census in two weeks with 10,000 enumerators in the field, tight planning and deployment management are essential. Even after careful estimation of the time required to complete each survey and providing each enumerator with a quota to complete, there will still be areas that cannot be completed in the required time and others that will be completed quicker.

A population census is a great example of a lot of data to capture in a short period of time. The addition of people and technology can increase efficiency, depending on the complexity of the task. If one person can complete 30 surveys in a day, scaling up the number of people can quickly increase the number of surveys completed.

In some locations, data capture may be relatively quick—everyone's home and willing to cooperate. In other locations, residents may be away at work or hesitant to communicate with enumerators until someone else in the household is present. Construction work may hinder access to properties, or a severe weather event—cyclone, fire, flood—may have picked the wrong time to hit. If fieldwork managers can see live fieldwork progress and some initial contact information from enumerators who weren't able to complete their surveys (for example, "Many people in this apartment block work shift work at the nearby factory—the switchover time is early evening, so come back then"), they can adjust schedules accordingly. They can move enumerators from completed areas to the shift-worker apartment block, so that the work can be completed in the early evening time slot when most residents are present.

Using dashboards and maps to present data on the fly when the fieldwork is a methodical observation of a place (vineyard), people (population census), or structures (water network assets) enables teams to track and assess the completion of data capture tasks.

Designing effective dashboards

The best dashboards are informative and clear. Consider the following elements when designing a dashboard that communicates real-time situational awareness.

Determine your audience

Who? | Important questions | Dashboard location

Avoid information overload

Necessary information | Images or videos | Too distracting?

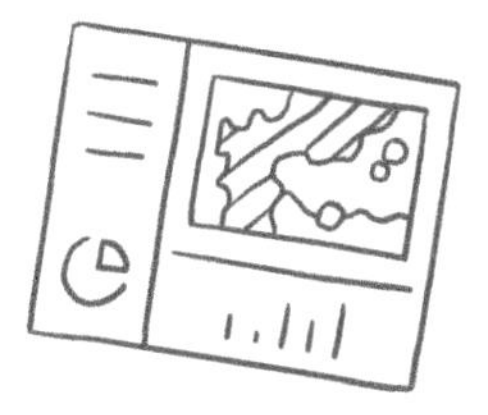

Provide context when needed

Types of values | Target or historical | Selection-based information filters | Criteria for automatic reformatting

Make good design choices

Organization | Size | Color

Changing your data capture method or tool while in the field

For all the excellent planning you do before you or your team go out in the field, there will be exceptions. Geography, infrastructure, weather, nature, and technology can all change. Always have a plan B, and when the stakes are high, consider a plan C.

Plan B may be a free-form set of GIS data layers, a spare device, and hot swappable spare batteries. Plan C may be a paper notebook and a camera.

If you are flying over the jungles of Papua New Guinea in a helicopter using expensive and complex measuring equipment, how you optimally cover the required geographic extent will need to adapt to ever-changing environmental conditions. And you must be able to record where you went—and perhaps where you didn't go—so you can come back to the locations you missed when the weather improves.

When performing fieldwork for scientific research as opposed to engineering design and maintenance, you should account for a certain amount of data capture freedom in your fieldwork. It's in the title, after all—it's research—you don't know everything about what you're researching. Semi-structured research questions, or in the case of fieldwork, semistructured data capture tools, can be of great use to capture information in the field to take back to the office for further analysis.

The simplest semistructured data capture tool is a digital camera. Photographs taken in the field are dated, provide context, and can be further analyzed on return to the office or sent to someone while fieldwork continues for parallel research that can guide ongoing fieldwork.

Taking semistructured data capture to the next level, consider having preplanned data layers available in your field GIS, in which features can be captured and incorporated into your organization's GIS at a later date. These layers are a significantly more sophisticated version of a map you might scribble on the back of the national park entry permit salvaged from your backpack.

On the topic of scribbling on the closest piece of paper at hand, let's return to the case of the paper notebook. No matter how digital your fieldwork project has become, there will inevitably be a need for a note, a reminder, a placeholder to be captured. A well-designed digital information system will have a place for these notes, of course. But with batteries dying, devices dropping, or the wrong cable being forced into the wrong port, only to render data transfer useless until a return to the office, having a paper notebook on hand still can't be recommended enough. Taking the step from scrap paper to a bound notebook alone is a significant evolution in field note-taking. Using the notebook pages in sequence and even dating the pages will have you well on the way to not just a plan B but a rich complementary tool that works with your field GIS.

The field notebooks of Charles Darwin and Maria Sibylla Merian have demonstrated the fine detail that can be captured with pencil, pen, and a few watercolors, far surpassing the detail captured in even the highest-resolution photography. For instance, a series of sketches that focus only on a particular body part—in the case of Darwin, the beaks of finches—can show the variations that provide information to support further scientific investigation. A side-by-side sketch of the male and female of the species can immediately demonstrate a size difference that is hard to see when you need to take separate photos of animals that refuse to pose neatly side by side for you.

Times have changed, and advances in digital photography mean that having powerful photo equipment on hand is within most fieldworkers' reach, but drawing in the field still has advantages. Even a simple annotated sketch of the base camp can quickly identify key locations to communicate to others.

It's also useful to consolidate your paper and digital records. Taking a quick photo of your sketch (with your device and app of choice) and attaching that photo to your field GIS record means that the sketch is readily accessible to anyone using the database (when synchronized). For the time you are in the field, your notebook is likely to be your preferred quick reference to all information about your fieldwork. Months and years later, however, when your focus is the data in the database, when you come across that image in the GIS, you should be able to trace it back to the physical notebook (at best on a bookshelf, but possibly in a box somewhere) and reimmerse yourself in the world of your fieldwork.

Going into the field is a costly exercise; returning to the office or returning to town where you can buy replacements from a store may not be an option. In these cases, be sure to have your plan A (first-choice hardware), plan B (duplicate first-choice hardware or secondary-choice hardware), and plan C (paper notebook) at hand.

Reconciling field data

You've designed your database well; you've carefully chosen unique identifiers so that multiple people can synchronize their data in the field at any time; internet connectivity is solid; and your fieldworkers can successfully send their information and receive updates from the central database to continue working. Welcome to field GIS heaven!

If you are not in field GIS heaven, you will need to reconcile field data records. Someone's device will break, lines will self-intersect and won't be able to be synchronized, or connectivity will fail. None of these will cause project failure but only opportunities for great detective work to ensue.

GIS data synchronization has been the most significant advance in fieldwork in recent years. When it works, we all have reason to rejoice. When it doesn't, many a smartphone has been sworn at and threatened with being relegated to the bottom drawer and replaced by good, old reliable paper. The most common reasons for synchronization failure are the following:

- No network connectivity
- Insufficient database permissions
- Invalid data

Network connectivity would probably top the list of most frequently encountered issues, but it also is usually the least disruptive. If you have chosen an app that can hold on to the data locally, when you next move to an area with network connectivity, you can synchronize.

Fieldwork testing should usually iron out issues with insufficient database permissions. Having fieldworkers sign in and perform a test sync should be able to catch this requirement early. When using a shared database, though, it is important to be aware of changes and upgrades being performed, as user permissions can be unexpectedly revoked without someone understanding the potential domino effect.

Invalid data may be the least frequent of these issues but probably produces the most frustration. A well-designed map, app, or form can minimize the creation of invalid data. Rules can be put in place that prevent a line from self-intersecting, prevent a value from being entered into a field outside a nominated range, or restrict the number of points that can be captured within the bounds of a polygon. Using such rules is the best strategy for reducing the frequency of issues with invalid data. But invalid data can still come from unexpected sources.

When a fieldwork project is handed off from one person to another, a reasonable place for the new worker to pick up from is the field GIS. When you inherit a project, be sure to also seek out any accompanying data, such as field notebooks or photos, that can be used to validate the information in the GIS.

I don't see the satellites you are seeing

NAME: Marika **ROLE:** Field app product engineer
INDUSTRY: Software development **CIRCA:** 2020

Invalid data can come from unexpected sources.

Recently, I was investigating an issue reported with one of the Esri field apps. The app was running on a laptop computer that was connected to an external GNSS receiver. It had been reported that no data was being captured in the app, even though the connection to receiver was clearly evident. The person who reported the issue was in North America, and I was in Australia.

I followed the steps that were provided exactly, and at different times of the day, outside our office and outside my home. I couldn't reproduce the problem. After some communications back and forth, my now GNSS buddy sent me an NMEA log file from his computer that I could play back on mine to simulate the behavior. Voilà! Instant fail. So, what was different? Well, it turned out that in Wisconsin, USA, my GNSS buddy could see a BeiDou satellite that I couldn't see in Melbourne, Australia. And because this app didn't know what to do with the BeiDou sentences, which were evident in the NMEA log file, it made a mess of reading the entire stream of data.

With some clever scripting, we were able to provide my GNSS buddy with a configuration file that would allow him to turn off the receipt of those BeiDou sentences, and the app was able to read all other sentences as normal. At that time, a complicated rewrite and republishing of the app was avoided, and a speedy fix was made available. As a bonus, we now had a configuration method that could be used to tailor exactly what sentences could be received by the app on any device with any receiver.

As I discovered, understanding the difference between what data is received and what data is expected by your apps is a critical step in reconciling data.

Notes

Afterword

What's next, and further reading

As you read the contents of this book—stories that span more than 20 years of fieldwork—you may ask yourself, is this still relevant? Surely, all the technical issues have been overcome, and isn't AI going to do all this for me anyway?

I wrote this epilogue on the closing day of the 2023 Esri User Conference in San Diego, California. Nearly 18,000 people had just come together to share their stories, curiosity, and excitement for GIS. The annual Esri User Conference is an amazing event and a must-see, must visit for anyone with an interest in GIS. The plenary alone is like nothing you can imagine: the scale, the emotional response by every participant, and the continual positive reinforcement that your work matters.

During the 2023 conference, my colleagues and I spoke to hundreds of fieldworkers, fieldwork coordinators, scientists, and engineers. We heard the very same types of stories that you have read in this book, over and over. Some people are still converting the paper data capture processes to digital. Some have complex digital methodologies already but know they can do more and came to quiz us on what else they can do. Many came with their notebooks, their mobile devices and computers, and fired up their projects in our showcase area for on-the-spot troubleshooting. All of them were excited to show us what they had done already and wanted to extend their use of GIS in the field even further.

You can read about current fieldwork stories and more on Esri's Field Operations website at esri.com. On the website, search for stories on the subject areas described in this book or by your industry. In these stories, you will see how people have solved problems and how organizations just like yours have used field GIS in their projects. You will see many apps and devices used. There is no one perfect combination that suits every need, but many great solutions exist that you can learn and choose from.

By keeping notes on your own fieldwork experiences in this handbook, you will build an invaluable resource for your future field GIS workflows. For every story that stays the same, there is another golden opportunity for improvement in the future.

Index

G

H

I

J

L

M

N

P

R

About Esri Press

Esri Press is an American book publisher and part of Esri, the global leader in geographic information system (GIS) software, location intelligence, and mapping. Since 1969, Esri has supported customers with geographic science and geospatial analytics, what we call The Science of Where®. We take a geographic approach to problem-solving, brought to life by modern GIS technology, and are committed to using science and technology to build a sustainable world.

At Esri Press, our mission is to inform, inspire, and teach professionals, students, educators, and the public about GIS by developing print and digital publications. Our goal is to increase the adoption of ArcGIS and to support the vision and brand of Esri. We strive to be the leader in publishing great GIS books, and we are dedicated to improving the work and lives of our global community of users, authors, and colleagues.

Acquisitions

Stacy Krieg
Claudia Naber
Alycia Tornetta
Craig Carpenter
Jenefer Shute

Editorial

Carolyn Schatz
Mark Henry
David Oberman

Production

Monica McGregor
Victoria Roberts

Sales & Marketing

Eric Kettunen
Sasha Gallardo
Beth Bauler

Contributors

Christian Harder
Matt Artz
Keith Mann

Business

Catherine Ortiz
Jon Carter
Jason Childs

Related titles

Top 20 Essential Skills for ArcGIS Pro

Bonnie Shrewsbury & Barry Waite

9781589487505

Mapping America's National Parks: Preserving Our Natural and Cultural Treasures

U.S. National Park Service

9781589485464

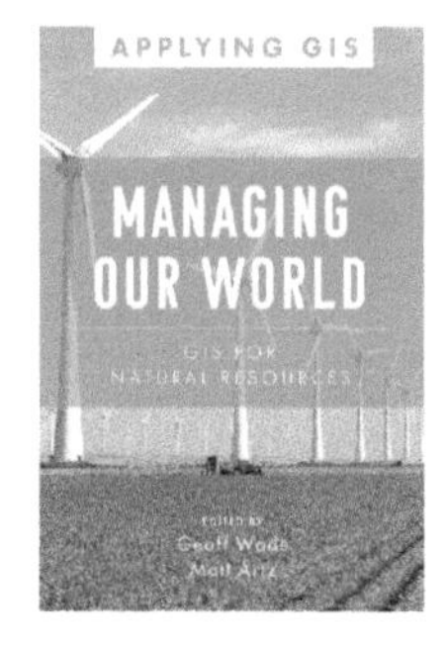

Managing Our World: GIS for Natural Resources

Geoff Wade & Matt Artz (eds.)

9781589486881

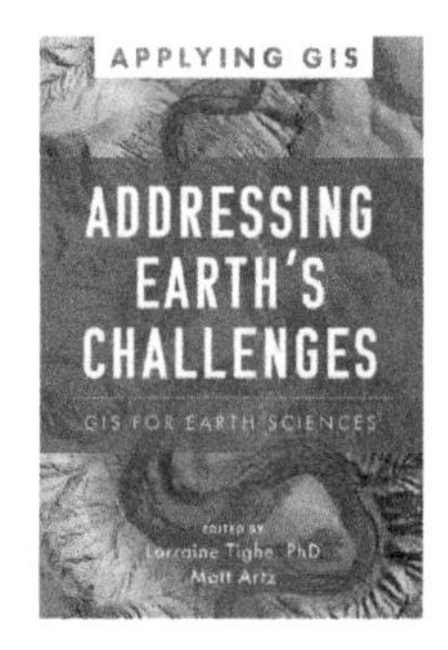

Addressing Earth's Challenges: GIS for Earth Sciences

Lorraine Tighe & Matt Artz (eds.)

9781589487529

For information on Esri Press books, e-books, and resources, visit our website at

esripress.com.